# AMATEUR RADIO, SUPER HOBBY!

## McGraw-Hill/VTX Series

*Calter*
Practical Math for Electricity and Electronics

*Jackson*
Practical Housebuilding for Practically Everyone

*Luciani*
Amateur Radio, Super Hobby!

*Wells*
Building Stereo Speakers

# AMATEUR RADIO, SUPER HOBBY!

## VINCE LUCIANI, K2vj

**McGraw-Hill Book Company**

New York  St. Louis  San Francisco  Auckland  Bogotá
Guatemala  Hamburg  Johannesburg  Lisbon  London  Madrid
Mexico  Montreal  New Delhi  Panama  Paris  San Juan
São Paulo  Singapore  Sydney  Tokyo  Toronto

Amateur Radio, Super Hobby!

Copyright © 1984, 1980, Vincent J. Luciani. All rights reserved.
Printed in the United States of America. Except as permitted
under the United States Copyright Act of 1976, no part of this
publication may be reproduced or distributed in any form or by
any means, or stored in a data base or retrieval system, without
the prior written permission of the publisher.

McGraw-Hill/VTX Series.

1 2 3 4 5 6 7 8 9 0 D O C D O C 8 9 3 2 1 0 9 8 7 6 5 4

ISBN 0-07-038959-4

**Library of Congress Cataloging in Publication Data**

Luciani, V. J.
  Amateur radio, super hobby!
  (McGraw-Hill/VTX series)
  1. Amateur radio stations.   I. Title.
TK9956.L76   1984      621.3841′66      83-25544
ISBN 0-07-038959-4


Book design by Grace Markman

**To my son, Jim, WA2JNN**

# WA2JNN JIM

| A.R.S. | DATE | GMT | RST | MHz | MODE |
|---|---|---|---|---|---|
|  |  |  |  |  |  |

COMMENTS : _____

_____

_____

WAS   WAC   TAD   RCC   ARPSC

RD1  P.O. BOX 111H
EGG HARBOR, N.J.
USA        08215

# · CONTENTS ·

# · PREFACE ·

> There is something about this book I like. It is written on a level *for* beginners. The author doesn't *say* he's writing for beginners, then, in a few pages, go far over their heads with the subject.

When published author and reviewer Charlie Buffington* (now deceased) reviewed the early draft of this book, the preceding quote was his opening comment; he also added many helpful suggestions, all of which were incorporated almost as though Charlie were watching over my shoulder all the while.

Charlie was right, of course. This easy-going book is for anyone and everyone to enjoy, and more. We in the hobby of amateur radio are keenly aware of what you outside the hobby are missing. We know that as soon as you lose some imagined fears of what amateur radio is and how to go about joining, you are going to ask, "Amateur radio, where have you been all my life?" And why am I so sure of this? Because the same situation is being repeated thousands of times each year. It's that simple.

It would have been easy for me to have written a book on amateur radio to impress you with my expertise. But had I done so, I would never have merited Charlie's keen insight into the true nature of this unusual book. Instead, I followed a far more challenging course by writing the first book of its kind on the amateur hobby, one in which I *say* I am writing for beginners, and then, in a few pages, do *not* go over their heads.

If you give this book half a chance, it can favorably affect your life no matter who you are, and in ways you couldn't imagine. And if you are considering a course in amateur radio

*\**Your First Personal Computer: How to Buy and Use It*, McGraw-Hill, 1983.

ix

studies, reading this book first will make your studies much easier in that you will have a better understanding of what and why you must learn—better motivated is what you will be. After all, what in the world do we ever learn best than that which we enjoy learning? It's that simple.

Nothing could please me more than to know, at some future time, that this book has been responsible for improving the lives of many people. Amateur radio can and has done much for people throughout the country and around the world.

Perhaps you will count how many times in the book I have promised you success in amateur radio and in all that goes with the hobby if you will only develop and hold on to two qualities—interest and desire. With these, everything else will fall into place. It's that simple.

I think Charlie would have agreed.

Vincent J. Luciani

## · 1 ·

# Getting Acquainted

If this book was given to you by a radio amateur friend who appreciates you enough to hope you will join him or her in our fascinating hobby, you are quite fortunate. You can soon be on your way to becoming a licensed amateur radio operator because you have the fine advantage of a friend who is ready to help you.

If, however, you bought this book on your own, perhaps because you have always had a personal interest in or curiosity about the hobby, then allow me to congratulate you. You too are on the right track toward soon becoming a licensed radio amateur. Through my book, I hope to be the friend you need to help you get started in this exciting hobby.

Either way, the door to a totally new, delightful, and hard-to-imagine world—the world of amateur radio—is now about to open for you. In this book, I will tell you not only about some of the many benefits and features of our hobby but also how you should go about getting an amateur radio license.

Along with this book, however, you will need two personal qualities. You will need to have an interest in the hobby, and you will need to have the desire to become a part of it. Interest

1

and desire are what it takes for the hobby to soon be yours to enjoy.

I have helped many people who once thought amateur radio was only for engineers. Years ago it used to be. It isn't anymore, but people still have hangups about what they think amateur radio is, and every time I think I've heard them all, a new hangup comes along.

People with fixed notions of things remind me of one of my favorite little stories which I'd like to tell you, not only to break the ice but also to prove a point that things are not always as they seem.

> Little Arthur was only 6. One day while sitting in his chair thumbing through some pictures in a book, he suddenly broke into tears. His mother saw that the cause for Arthur's concern was a painting of Christians being thrown to the lions. "Mommy," wailed little Arthur, "that poor lion over there doesn't have anyone!"

So you thought you had an idea of how that story would end, did you? Are you as sure of what this book is going to be like in describing amateur radio? Things are not always as they seem.

Later in this book I will explain in more detail exactly why the amateur license is not the problem it seems to be. For now, however, let me say only that we have what is called a *novice class license* for newcomers which turns out to be as easy as pie to get. Really.

You will also read about an organization for the physically handicapped which helps them get into the hobby and even loans them specially modified equipment. Amateur radio is a great equalizer for the handicapped because on the air everyone is equal.

Let me also hurry to tell you what this book is *not*. It is not a technical manual. This book does not teach the technical side of radio, although, by the time you have read it through, you will have learned enough about the "inside" of our hobby to be able to confidently converse with other radio amateurs in their own language. In fact, you may one day be surprised to find out that your pleasant reading of this book actually did teach you enough to pass about half of the novice class amateur radio examination. You'll know better what I mean when you get to the end of the book.

To be perfectly honest, though, I must admit that my book does do some teaching, but not in quite the way most other books do. I have written this book in an unusual way because I've found that it takes something different to convince people of what I already know but they don't—that getting into amateur radio is truly so very much easier than it used to be and that you don't need a college degree to get into the hobby. Honest, all you need is the interest and desire. The basic words and descriptions in this book will show you that it really isn't so difficult to learn what amateur radio is all about. So relax— you won't be running into any of the high-powered technical terms that too many of the other books use in "basic" amateur radio.

The truth is, this informally written book is for those who have a hankering to learn what we radio amateurs are all about. I will give you a quick look at the hobby from my personal point of view. But not all that I've written comes from my own experience. A few of the pleasant surprises inside this book are the stories I have obtained from personal interviews or correspondence with some special amateur radio operators. Some of their names you will recognize: others, you won't. All of them talk about their experiences as amateur radio operators. Why should you read about them? Well, these people have a message, in their own words, that will be useful to you along the trail toward the amateur license. I suggest you pay attention to each story, and it doesn't matter that you don't know the individual. What does matter is how they succeeded in the hobby because if they did, so can you.

One day, when you are licensed, you will find out for yourself that the hobby of amateur radio is super fun, but the fun of operating the radio is only one part of it. The people are wonderfully different too. Times and people change, but so much of what was all good went into this hobby from its very beginning that no force on earth could ever take away the inherited spirit of amateur radio.

I have always been infected with an enthusiasm for amateur radio, an enthusiasm that allows me to write to you in this way. I sincerely hope that my "infection" will be passed on to you so that one day you too are operating your radio on the amateur bands and, perhaps, we will talk together. And if we do, please tell me how you got started in the hobby, and what the hobby has since done for you to make your life better. I know you will have much to say about this. Everyone in the hobby does.

# · 2 ·

# What Is It?

What is amateur radio? Sorry, there is no simple and clear-cut answer to the question. Anyone who claims to have such an answer in a single statement is likely to be either a college professor or a government expert or someone else who feels they must "solve" everything they face. It just isn't that way in amateur radio.

I could say, for example, that amateur radio is a hobby. That's certainly true. I could say it is the ultimate hobby, but that would be a prejudiced claim, no matter my experiences or the experiences of others.

I can also claim that amateur radio is much more than a hobby. Well, it is. As a hobby, amateur radio is a hundred hobbies in one because there are at least that many different ways to enjoy amateur radio.

Amateur radio is certainly not like some other hobbies which are easy to define and describe. Truthfully, I just don't know how to go about answering the question of what amateur radio really is. At least I can't answer it in a few sentences. Perhaps, somewhere in this book, each of you will find your own answers. In fact, if you have the interest and desire, I can guarantee you will find your own answers to the question. But these will not be answers so much as they will be trails that will sooner or later lead to the answer. Lucky you if you stay along those trails.

In asking other people about their impressions of amateur radio, most had the same thing to say: "You get to talk to countries all over the world."

Yes, we do get to talk to countries all over the world, including countries I am sure you never heard of before. (That is one hidden benefit to amateur radio for youngsters—they get to be geography whizzes in the most exciting and unforgettable ways, talking to or trying to talk to faraway countries.)

But hold on. Talking to these distant lands is only a small part of amateur radio. While you can certainly do as much talking to other countries as you want, if talking to other countries is enough to bring you into this hobby, then you're going to find a ton of extra fun because there's so much more to do.

Talking to other countries is really only a fraction of the kinds and varieties of activities we get into through our hobby. There are loads of things to be experienced. Multiply by ten, twice, an enthusiasm for talking to foreign countries and you only begin to get close to the hobby's hidden value as a "companion with a personality."

Guess what? If I were really backed into a corner to give a definition of amateur radio, I would say that I just did so in a sneaky way—companionship and personality. Think of it this way: The equipment gets to be your means to the conversations which provide the companionship of people all over the world, all over the country, from your own state, from your own town, and often from your own neighborhood. And with so much of what we do in amateur radio being so unusual, so different from other hobbies, we also find a personality built right into the hobby.

That's how it is in amateur radio—companionship and personality. If you're a lonely sort, or a shut-in, please stay with it to let this amazing hobby enter your life. You will never regret it, believe me.

In the chapters to follow, I'm going to talk about some features of the amateur hobby and about how you can go about getting started toward the license. I already explained that this is not a technical book and that it does not teach radio theory, so don't expect too much. What you can expect, however, is an understanding of the radio bands we operate our sets on, the different classes of amateur licenses that are available, and the operating privileges each has to offer.

I'll also discuss the super sport of talking to foreign countries (*DXing,* we call it—a term I'll later explain although it, and all other "technical" words I use, are explained in the glossary).

And I'll talk about a few of the hundreds of contests and operating activities we have on the air. I will also describe how some members decide to become active in sending traffic messages like they used to in the "old days" when all this started. Sending messages is fun. Many of us get into it, but I want you to understand right now that there is no requirement for you to operate traffic nor is there any operating you *must* do. Everything you do in amateur radio is of your own choice, within the rules and regulations, of course.

A lot of people tell me amateur radio is too expensive for them, but they don't know what the costs really are. You can, for example, actually put together a novice station for well under $100. I'll discuss equipment costs in a later chapter. I'll also tell you about the various amateur radio magazines, and which of them might be best for you.

Tucked away at the very end of the book is a little true-false quiz for you to take. No one says you have to; it's up to you. My reason for your taking it is explained back there. You may be surprised at what you will learn from an evening's entertaining reading on what used to be thought of as a difficult subject. It's all in how it's done, so that it becomes understandable and doesn't frighten you away with technical mumbo-jumbo.

As you work your way along this book, chapter by chapter, you may or may not realize that I'll be hard at work in trying to convince you that you can make the grade in our hobby, and I will try even harder for shut-ins and handicapped.

And after describing some of the benefits to the hobby (there are too many to ever cover them all in one book), I will go about the description of exactly how you should take your first moves toward joining. I have all sorts of information for you on this subject, whether you care to take one of the free courses around the country or whether you want to do it on your own, at home. Either way, you may be pleasantly surprised to learn that we "inside the hobby" are always looking for new members. We're always anxious to help others join. If you follow my advice in the later chapters, you will very likely be assured of a radio license.

With all those subjects, and more, you can be sure there are several other subjects I've had to leave out so that I could keep your attention. And it is important that I keep your attention because every chapter in this book was selected for good reason. Every chapter will be useful to your future in the hobby. Trust me.

But if you doubt my word, then check yourself out the next time you're at the bookstore as you pick up any nonfiction book with photographs in it. What do we always do? We flip through, looking only at the pictures. It's human nature to do that. But after all, the author really put a lot of thought and effort into the words in the book, and that's what you should be judging (unless it's a book on photography, or such).

Well, I've used words, and a few amusing drawings by artist and radio amateur Joe Chistolini, to paint the story of amateur radio so that I could lure you into this hobby. My words are the "pictures" by which I hope to both generate your interest and to form your desire to join us.

# · 3 ·

# Tinkerers

Along with the privilege of operating a radio station, the amateur radio license also gives you the right to tune, repair, or even build your own radio equipment for use on the air. In fact, everyone in amateur radio is encouraged to work on their own equipment and to tinker with it. Your only limit in doing so will be the limits of your own new talents.

When you pass the amateur radio examination, you will have demonstrated that you have the qualifications to operate and adjust your own radio set. In time, you may wish to learn more about radio so that you can design and build some equipment of your own. In a later chapter I will talk further about building your own equipment. Quite a few newcomers build a little gadget or two for their radio stations, and these are the people who are likely to be anyone such as retired cops or insurance agents who never would have believed a few years earlier that they had it in them. What they don't know, and what you don't yet know but will be suspecting after a dozen or so chapters, is that something in amateur radio rubs off on everyone.

Tinkerers, tinkerers. Throughout the several decades since radio was first developed back around the turn of the century,

9

radio amateurs had been (and still are) known as tinkerers. Experimenters.

The ancient equipment used back in the earliest days of radio long before the vacuum tube and transistor were invented, were marvelous contraptions that required its owner to be a chemist, plumber, and carpenter and to have a bunch of other skills in addition to radio. It took these extra skills to get on the air because, back then, there were no radio parts stores over at the shopping mall to run to. They did it all, back then, even to building their own parts in many cases. They had to. They were true tinkerers.

Radio amateur old-timers from the first few decades of the century like to talk about how such ordinary things as Quaker Oats containers were among the radio amateur's best friends. In fact, probably more Quaker Oats containers were found in those old amateur radio sets throughout the country than in most kitchens. That's because the Quaker Oats container was just the right size around which to wrap coils of wire for the radio.

We may never know who the very first person was to receive an amateur radio license, but two old old-timers I can tell you about are Art Ericson, W1NF (his present amateur radio call sign), and Irving Vermilyea, W1ZE. Ericson's start goes back to 1902 when, at the age of 7, he built his own radio set from parts that came from a druggist, a boat yard, a chauffeur, and a hardware store. Vermilyea was into wire telegraphy in 1901 and very soon afterward made the change to wireless telegraphy.

I could go on about how those old-timers worked through one building problem after another with their radios, using gadgets they found in junkyards, bought at hardware stores, or made for themselves. (Of course, I know about these things from hearsay; I'm not that old.)

Indeed, much can be said for the experimental, tinkering nature of those inquisitive radio amateurs of old. Tinkering was the radio amateur's personal expression of the art of radio, and amateurs contributed much to the scientific breakthroughs of the times because they were constantly on the alert, constantly advancing into unexplored fields. They had to. It was nearly a matter of survival in those first years of the hobby because every time the amateurs produced a scientific break-through or proved that something previously thought to be impossible was not only possible but practical and worthwhile, that was when the commercial companies would move in and take over from the amateurs. It may seem like a greedy op-eration on the part of the commercial interests, and in a way it was, but it all had its net good effect because the amateurs simply went ahead and developed something better. Such tin-kering advanced the radio communications art in ways seldom appreciated, then or now, by the public.

And even today, we tinkering radio amateurs are likely to buy a new piece of equipment and, in short time, have it on the workbench for a modification intended to make it work better, at least for our own uses.

It is true that we don't build as much of our own sets as we used to but that's because it has all gotten quite a bit more complicated (and they make radio parts so small nowadays that I need a magnifying glass to see them). But we will always have among us those who are ever eager to experiment with some-thing that they either read about in one of our technical mag-azines (which are described in a later chapter) or designed themselves. This, you see, is the true tinkering tradition of amateur radio which comes with the territory, and we all get that way. You will too, to some degree.

It all reminds me of another story I'd like to tell, this one about the tinkering radio amateur. The story goes like this:

In a faraway country there was a preacher, a lawyer, and a radio amateur who had been convicted of a crime against the country. Without wasting any time, the dictator ordered them to be immediately executed by guillotine.

The condemned prisoners were led into the courtyard. As they approached the guillotine, the preacher asked to go first and to be placed over the base of the guillotine on his back so that he might spend his last moments on earth looking upward toward heaven. His wish was granted. As the huge blade was released and came hurtling downward, it stopped suddenly, only inches from the preacher's neck. A miracle! Impressed, the dictator took it as a sign from above and set the preacher free.

Next came the lawyer who, trained to respect precedent,

also asked to be beheaded facing upward. His wish was also granted. And, again, as the blade hurtled down, it once more stopped suddenly, only inches from the prisoner's neck. The dictator set the lawyer free also.

Then came the turn of the tinkering radio amateur who also asked to be placed on his back at the guillotine. And, as all was made ready to again release the blade one more time, the tinkerer shouted out, "Wait a minute—I think I see the problem."

And with that, such future anecdotes will be few and far between.

"I think I see what's wrong."

# · 4 ·
# Who Can Join?

Before I try to answer the question of who can join amateur radio, please bear with me while I also try to qualify two important notions about our hobby.

First, getting an amateur radio license takes more than simply filing an application. In order to get a license, you have to take a test on basic rules, regulations, and radio communications theory.

But, listen, just because you need to take a test to get the license shouldn't slow you down. You see, there is a special novice class license which serves, in a way, as apprenticeship to amateur radio.

Although the technical requirements for the novice license are almost nothing, by the same token, you wouldn't expect to earn the full bag of apples by having that novice license, would you? Of course not. The privileges of the novice class license are limited. The idea of it is that once you have had a taste of what amateur radio is, and of having found out for yourself how much more there is to sample in this fascinating hobby, the hope is that you will stay with it long enough to get one of the higher licenses. (Chapter 17 tells all about the different classes of amateur radio licenses.)

In this hobby you can go as far as your talents, your interests, and your desires take you. So keep cool, and aim for the novice license first. The rest comes, in time, and it comes more easily if you first get a taste of the hobby with the novice class license.

Second, we have a wide and nearly endless range of operating and nonoperating activities you can enjoy in amateur radio, even for the novice. This fascinating hobby does, indeed, seem to offer something for everyone.

It isn't at all necessary—nor is there ever a requirement—that you must do or join something or other in amateur radio. What you do is all at your own preference and by your own wishes. Keep that in mind, please.

What I might suggest is that you eventually develop whatever you find that you like best about the hobby. It is good to have special interests, particularly when there is such a wide range of interests available to you. Specializing allows you to be good at something in a hurry without spending a lifetime at it. There are so many features and benefits to amateur radio that you needn't worry about not finding any you like. We all find our own selection of preferred activities. It comes with the hobby.

Later I'll get around to talking about some of these benefits and activities, but no amount of writing could ever list all of them. Besides, many of them are best experienced rather than read about. You'll get to understand what I mean as you go along.

In going back to the question of who can join, this is a good place for me to assure you, again, that in no way is amateur radio a hobby of professional experts. Absolutely not. In fact, these days only about 1 member in 10 has ever had any radio background before becoming a radio amateur. Look over this list of radio amateur's professions (this list is from my chess and amateur radio membership list which you will read about in a later chapter): student, attorney, hairdresser, housewife, clergy, insurance agent, artist, musician, teacher, construction worker, magician, accountant, psychologist, machinist, biochemist, bonded courier, retiree, surveyor, pilot, meteorologist, pharmacist, clerk, writer, physician, carpenter, chemist, draftsman, steelworker, architect, and so on. Impressive, is it not? Very few electrical engineers in the lot.

Age-wise, that same chess and radio membership list has one member aged 9 and another aged 77. That's nothing—not too long ago an 87-year-old passed the novice exam. So did a five-year-old, which you will read about in a later chapter. Can you imagine a kindergarten kid already licensed as a radio amateur? Show-and-tell must be something else in his class when he comes in bragging about talking to this or that country the night before.

For your interest, and to follow through with more information in answer to the title question, the typical percentage breakdown of U.S. radio amateurs (of which there are nearly 450,000) in age groups is: under age 20, 9 percent; 20–30, 13 percent; 30–40, 23 percent; 40–50, 17 percent; and over age 50, 38 percent.

Those people over age 50 aren't just old-time members to the hobby. Not at all. In fact, probably the fastest-growing group in our hobby are the retirees, those who learn in a hurry all the special benefits that amateur radio offers including talking free around the country to friends and relatives, even with those friends who are not radio amateurs. (You didn't know we could do that?)

I've been talking about how just about anyone can pass the novice test, and you are probably wondering how much technical knowledge one needs to be able to pass this test. Well, if you can divide two numbers and know how to apply the answer, that's about as much math as you are going to need for the novice test. That's a promise. I'll give you a sample of novice radio theory in a later chapter, and a pretty good idea of how "complicated" the novice test isn't.

There's another joy to taking the novice exam I'm sure you will appreciate. You don't have to drive to the "big city" to take this radio exam. Nope, the novice test is given by any qualified radio amateur who is above the novice class and who is not a relative to you. Most radio amateurs have given someone the novice test at one time or another, often right in their homes. I've done it several times. You can't beat the convenience, can you? It's just another example of how they've made it so much easier, these days, to join amateur radio.

# · 5 ·

# Then, and Now

Amateur radio had its start at the turn of the century, right on the heels of radio's discovery in 1896. And by 1914, the hobby was already in full stride.

There were few radios of any kind around in 1914, whether for amateur use or commercial. In fact, amateur radio operators were not licensed in the beginning simply because the whole thing was so new that no one knew what to make of it. Not until 1934 did the Federal Communications Commission (FCC) set up shop as the government force to regulate things.

In the early days of radio, communicating great distances meant talking to someone in the next state. Being experimenters, amateur radio operators soon got the urge for greater distances and began to develop their own equipment by which to do so. They also improved on their operating techniques in order to span the country from coast to coast by relaying messages from one to the other.

It may not sound impressive to we modern people who turn on a pocket-sized transistor radio to hear shortwave broadcasts from Europe, but you must consider those days of old when radio was a brand-spanking-new baby. Most people had never even heard of it in 1914, so getting a message from New

York to California by radio was quite exciting for the radio amateur of that day, and messages crossed the continent through organized relay teams.

From this early practice of relaying messages by radio came the organization now known to all radio amateurs—The American Radio Relay League (ARRL)—cofounded in 1914 by its first president, Hiram Percy Maxim. Several decades later, the ARRL is still the organization that leads the nation's nearly 450,000 radio amateurs.

Several thousand amateurs from the United States and many other countries annually visit ARRL headquarters. My son Jim and I did in 1980; it is a pilgrimage which all radio amateurs try for, at least once in their lifetimes.

The ARRL facility is a large, modern two-story building that is bustling with activity. Still, everyone we talked to was most courteous and willing to spend time with us. In fact, while I was there they issued the Old Timer's Certificate to me, on the spot, which I thought was a nice memento of the trip. (This award is earned by being in the hobby 20 years or more.)

The original headquarters building was moved from its former site to the new site, where it now houses many ancient records of amateur radio, plus the ARRL radio station W1AW, which we operated while there. (W1AW is the station's radio call sign—you will read about call signs in a later chapter.)

I found it interesting that Connecticut's vacation guide listed the ARRL site as: "Headquarters of worldwide ham radio organization. Maxim Memorial Museum has 19th, 20th century communications aids." Maxim's original rotary spark gap transmitter (see glossary) is still there, in working order. (For a copy of the Connecticut guide, call 1-800-243-1685.)

The ARRL communicates to its members through the publication *QST* and through regularly scheduled broadcasts on several of the amateur bands via its commemorative station W1AW (which was Maxim's radio call sign). *QST* is available in Braille and on talking tapes.

*QST* also serves as the official publication of the International Amateur Radio Union (IARU), an organization for radio amateurs throughout the world. (A complete list of IARU member organizations is included in the appendix.)

(There are also several other national radio amateur publications available which are described in a later chapter, and there are over 50 of these, perhaps many more, published throughout the world by the various amateur societies.)

Technical development in radio communications is an important part of amateur radio. Amateurs were largely responsible for the progress of radio through the 1920s. As radio amateurs of the period would successfully show the usefulness of certain radio bands, pressure would come from the commercial interests who would want them for their own uses and would try to move the amateurs out. (Amateur radio bands are explained in a later chapter.)

This competition for radio bands was not all bad, however. As a result of the pressures to push amateurs out of the developed bands, they moved out into the unexplored regions of radio in order to keep their hobby alive. By doing so, the amateurs made much technical progress simply because, on their own, they would soon develop and adapt to the newer and higher bands that were previously considered worthless. That's how it was in the very early days of radio.

Today? Would you believe that radio amateurs own and operate amateur television studios? In fact, radio amateurs were already experimenting in amateur television broadcasts back in the 1930s, long before the "tube" got to be a mass gadget for every household. Today there are amateur television relay stations (repeaters—described in a later chapter) that automatically relay television signals from a remote station.

And computers? My goodness, computers are now everywhere in amateur radio, both in our operating and in our equipment. Groups of amateurs meet regularly on the air in organized get-togethers to discuss their own brands of computers. There are dozens of "nets," as we call these gatherings, for each and every computer ever made, it seems. Those who take part in these computer gatherings are a fine group of experts who work together for one of the hobby's greatest traditions—technical progress.

They are now sending their own computer programs by radio. They load up the program, set the radio transmitter, flip the switch, and there it goes, over the air. Since the receiving station that deals in this form of amateur radio transmissions is likely to also have both a computer and printer, the program is duplicated at that station upon receipt by radio. I'll tell you, it all gets to be too much for me.

Coming up full circle from what was to what is, you might also be surprised to learn that we radio amateurs are out in space, and in more ways than one.

That's right—we've been in space since 1961 with our very own satellites circling the globe; we call them OSCAR, or Orbiting Satellite Carrying Amateur Radio. These satellites were designed and built by amateur radio operators from various countries. Our tenth, OSCAR-10, was launched in orbit in 1983, so you see we are quite serious about them. (Address of Amateur Radio Satellite Corp. is in the appendix.)

Many thousands of radio amateurs have specialized in satellite communications, and the prospects for the uses of our OSCARs grow with each launch to where, one day in the not too distant future, we expect we will have a wrist radio that will let us talk to amateurs in other countries via satellites.

Ours is not the only country to have amateur radio satellites. Russia has them, and Canada, France, and Japan plan launches of their own. In amateur radio, there are no international hangups—we use their satellites, they use ours. No problems.

And we've been in space another way. You see, astronaut Owen Garriott, known to us as radio amateur W5LFL, operated an amateur set aboard the space shuttle *Columbia* in 1983 to set another American "first" in space.

# · 6 ·

# Hobby Benefits

The amateur radio writer who is able to list and properly describe the full range of benefits that come with being an active amateur radio operator will also be the genius who has handled such simpler tasks as to have converted lead into gold, located the Fountain of Youth or built a working perpetual-motion machine.

My point, if I need make it more clear, is that it just can't be done.

We amateur radio writers make an effort to transfer our affections and pleasures with this fascinating hobby of amateur radio into words, but the very best any of us can do is to say that the benefits and joys of amateur radio are to be experienced, not just read about.

Admitting to a "product fault" is not very good advertising; therefore, I must meet the challenge of this chapter with the aid of spiritual inspiration if I am to make my point. But if I do not do so in this chapter, it is no loss. After all, throughout this entire book I have subtly suggested that few other activities on earth have quite the same worth in the lives of people as does amateur radio.

For example, let's talk about what amateur radio can do and has done for young people. One wonders how many of today's outstanding citizens, now amateur radio members, once skirted the fringe of the law, only to be completely turned around by an introduction and total dedication to this hobby. I can assure you that once the spark of interest is lit in youngsters, they soon take off into a new world, one that is ideally suited to their young spirits; soaring with pride over accomplishments and pride over learning new and exciting subjects, all of keen interest to them. And then the demand is fully upon their time and energies such that none is left over for mischief. This is an old story in amateur radio; it can be a new story for thousands more. For example, could there be amateur radio instruction in a detention home? You bet! Do those kids have the potential for learning amateur radio? Do eagles fly?

Those same youngsters have a habit of growing up. And when they do, they are concerned about jobs. Are jobs an amateur radio benefit? They certainly are!

Wayne Green, editor and publisher of *73 Magazine* (who is the subject of one of my later chapters), forever speaks out on youngsters in amateur radio. Here is what he once wrote: "We have enormous potential [in amateur radio]. We have one of the most fascinating hobbies there is; . . . the kid who gets involved with a high-technology hobby in school has a tremendous advantage. There are no bread lines for electronics or computer technicians, only thousands of jobs advertised in papers from coast to coast." Organizations which sponsor youth activities, take note!

And at the other end of the age scale are those who also gain a tremendous lift from amateur radio. I refer to retirees.

Whatever else one does in retirement, adding amateur radio should be a must-do thing. Honestly. Retirees from previously busy work lives who find retirement to be not sufficiently filled with things to do absolutely should look into amateur radio. In fact, our largest class of joiners, these days, seems to be the lucky retirees who had a friend in the hobby to lead them into it. Those that join make fine use of the hobby through daily contacts with friends all over the country. They also meet many new friends, keep productively active, and enjoy learning new things. They know that amateur radio is one of the best friends they have ever had in their lives.

Whether you are young or old, working or retired, let this

book form pictures of hobby benefits as you read through. Let it, for example, allow you to imagine yourself lazing around the deck of a yacht in the Atlantic Ocean, talking to someone in the Mojave Desert by amateur radio. Why this scene? I don't know, except that it might exercise your mind into considering the wide range of operating features we can get into. If this scene doesn't reach you, then how about the one that was written up in one of the amateur publications of the radio amateur who, while clinging to the edge of a cliff on Mount McKinley, Alaska, talked directly to his wife in Utah using a little amateur radio walkie-talkie. Fit anything you want in between those two scenes and you will still be unlikely to come up with something that some radio amateur, somewhere, sometime, has not already done.

As you read farther into this book, you will find your own answers to the question of what amateur radio can personally do for you. There will be instances, as you read, when you will think, "I didn't know that!" or "Now that's what I like!" Those will be the times when you will relate what we are doing in the hobby to what you might do when you too are a member.

Frankly, if each of the 450,000 of us were to list the 10 features of amateur radio we like best, our list of different items would be staggering. Which is why I go around calling amateur radio a super hobby. It is many hobbies rolled into one.

Relaxation is a key feature to amateur radio; we mostly agree on that point. True, this super hobby may not prevent ulcers, but I am convinced it can go a long way toward a cure.

One other thing: You're hardly ever lonely when you have an amateur radio set to operate.

# · 7 ·

# Amateur Bands

One of the first questions asked by many people who become interested in amateur radio is, "What radio bands do you operate on?"

*Radio bands* are places on the radio dial where different users are allowed to transmit. There are all kinds of radio users, these days, and you are already familiar with many of them, as I will soon show you. Some, for example, are the CB (citizens band) bands, the police radio bands, AM and FM broadcast bands, and TV bands.

I am going to give you a rather interesting explanation of radio bands, not only of the amateur radio users but also of many others who use the airwaves. Seeing how they all fit into the radio bands is a good way—perhaps the best way—to get an understanding of the amateur bands. In other words, if I tie it into what you are already familiar with, then it won't be hard to understand amateur bands.

All radio assignments come from the Federal Communications Commission and all users are assigned to certain bands. You can accept that. After all, if anyone were allowed to operate anywhere they please, there would be mass chaos and no one would know where to tune in anything. Therefore, all the various radio services are assigned to certain places on the dial. Amateur radio is no different from the other services—we are allowed to operate only on certain assigned bands.

Unlike many other services, amateur radio bands are scattered throughout what we call the *radio spectrum*. Let's take a minute to go over that term, spectrum. My dictionary defines spectrum as: "the band of colors formed when any form of energy is broken up; . . . in radio, the wavelength range of bands." In other words, a spectrum of radio waves is the whole collection of all radio bands put together. So when I use the term spectrum, I mean all the radio bands. When I talk about a *band*, I mean only one part of the spectrum.

First, you need to know that we have a name that goes with these bands. After all, a band is a spread of area on the spectrum; it isn't just one point. There is a term by which we say that the spread of an area in the band is between "so many of something or other." That "something" is what we call the *frequency*.

In radio work, frequency has a technical definition. My dictionary defines frequency as the number of periods or regularly happening events of any kind in a unit of time, usually in 1 second. This explanation is too technical for me, so I have come up with a simpler one, as follows:

Frequency is the number of times something happens, usually measured in a 1-second time period. In radio work we are talking about radio waves. Radio waves have their ups and downs, just like sea waves. We say they go up (positive) and down (negative). For those of you who have studied algebra, you will know that the drawing shown is of a *sine wave*. If you don't already know what a sine wave is, don't worry about it: it really isn't important to you at this time. Instead of calling it a sine wave, I am going to call it a radio wave; they are the same for the purpose of this explanation because radio waves are shaped like sine waves.

In this drawing of radio waves, let us say that the wave starts at zero time on the left of the drawing. At the beginning of the wave, it has a value of zero because it is just starting out. This zero value is sort of like the sea when it is calm and has no waves.

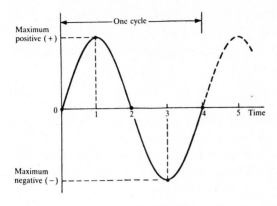

Ordinary sine wave.

At time 1 in our drawing, the wave is at its peak positive value. At time 2 the wave has returned to zero level. At time 3 it has gone down to its most negative value. At time 4 it has come back up to zero (where it started from at time 0). You can see that the radio wave goes from zero, up to a maximum positive value, back through zero on its way down to its maximum negative value, back up to zero, and on and on, repeating itself over and over again. That's how ocean waves are; that is exactly how radio waves act. They are cycles of radio energy that repeat themselves over and over again; up and down, up and down, always moving outward.

(If we were able to see radio waves going out over the air, they would look very much like a pond when a pebble has been tossed into it—waves continuously moving out from a center point, waves that have their ups and downs.)

If the time period we are talking about from time 0, in the drawing, to time 4 is 1 second, then we say that the frequency (that is, the number of times something happens) is one cycle per second.

But suppose the time period from time 0 to time 4 was one-thousandth (0.001) of a second. That would mean that we would have one thousand (1000) of these cycles happening in 1 second. We would say that the frequency of the wave, in that case, would be one thousand (1000) cycles per second. Okay so far?

Well, that is exactly how we identify where the amateur radio bands are located. We say they are on such-and-such frequency, and we give that frequency in cycles per second.

We don't really. We say *hertz* these days. Hertz (not the car rental) is the term now used for the unit of frequency. It means cycles per second. We use this term in honor of the German physicist Heinrich Hertz, who developed so much of the early theory of radio waves. But just remember that any time in radio work you see something marked as "hertz" (which is abbreviated Hz), you know that what it means is cycles per second, or frequency.

In technical work we use a shortcut for numbers as large as thousand and million. We use the Greek words instead. For example, for one thousand (1000) we use *kilo,* and I am sure that kilo is not a new term for you. You have heard of kilometers and kilograms before; they mean one thousand meters or one thousand grams. We do the same in radio; we use kilohertz and megahertz. Abbreviated, we write them as kHz for one thousand (1000) cycles (hertz) and MHz for one million (1,000,000) cyles (hertz). One thousand hertz is 1000 cycles, which we write as 1 kHz. One million hertz is 1,000,000 cycles, which we write as 1 MHz. Not so hard, is it?

To summarize, the radio spectrum is made up of all kinds of radio bands. Radio bands in turn are made up of collections of frequencies which are cycles of radio waves. Radio waves are measured in units of hertz. Kilohertz (kHz) is thousands of cycles per second; megahertz (MHz) is millions of cycles per second.

It isn't important to you at this time, but later on, in Chap. 18 on the novice license, you will learn about ac and dc, that is, alternating current and direct current. However, I want to mention here that radio waves are alternating current, ac. As I said, that fact isn't important now, but if you keep it in mind in that later chapter, you will get an extra benefit from it.

Now we can look at the big diagram with all those radio bands marked on them. Instead of looking at it as a complicated thing, all of which is new to you, how about thinking of this drawing in terms of its being a high-rise apartment building in which live all the radio spectrum users.

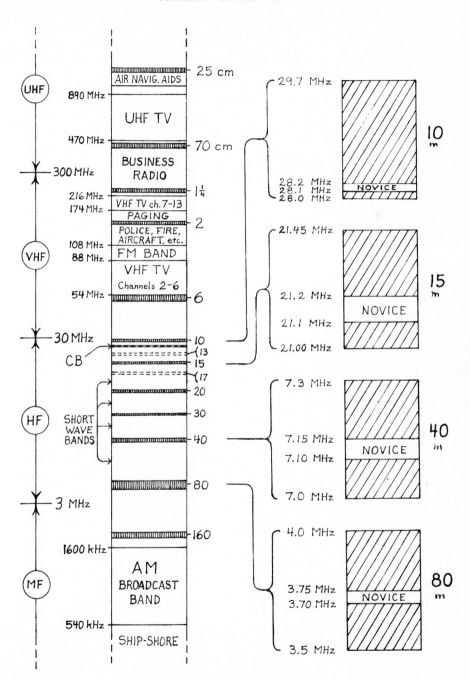

Radio and television bands.

On the ground floor of our Radio Spectrum Apartment Building are the tenants who use the ship-shore frequencies. These frequencies are below 500 kHz. (Remember: that is 500 kilohertz, or five hundred thousand cycles per second.)

On the next floor is one of the most familiar radio bands in the world, the AM broadcast band. Everyone knows about the AM broadcast band. Transistor radios tune the AM broadcast band, as do all car radios. The AM broadcast band has a frequency range from 540 kHz to 1600 kHz (540,000 cycles per second to 1,600,000 cycles per second).

You can see why we have abbreviations for these things, can't you? Why write out 1,600,000 cycles per second when we can simply abbreviate it to 1600 kHz or, even better, to 1.6 MHz? They all mean the same thing.

Next floor up in our apartment house are the more common amateur radio bands: 160 (meters); 80; 40; 30; 20; 17 (new band soon to be authorized for use); 15; 13 (another of the new bands); and 10. These are the only bands of interest to novice class radio amateurs, as I will explain in Chapter 18.

Let me talk a bit more about the rest of the radio spectrum, after which I will come back to the amateur radio bands.

In the space between the 160-m band and the 10-m band, there are many other users. The biggest users are the shortwave broadcasters. They are mixed in between the amateur bands. I'll come back to the shortwave bands in a little while.

You can see where the CB band is on the chart: 27 MHz. Above the CB band is the amateur radio 10-m band, and above that is the amateur radio 6-m band (not authorized for novices).

Next in our radio apartment building are the first few TV channels, channels 2 through 6. Living above them is another of the most familiar bands around, the FM broadcast band. Above that are various services that you find on your scanners—police, fire, aircraft, taxi, and so forth. Then we have another very popular amateur radio band called the 2-m band, which you will later read about.

Above the 2-m band are the remaining VHF TV channels, channels 7 through 13; then a few more less-used ham bands; various other services, including business radio; and UHF TV. Above UHF TV are many rather interesting services which you can learn about on another day; they include aircraft navigational aids, radars, NASA space frequencies, and a whole list of exciting users. But, as I said, those are for you to learn about on another day.

Now, about shortwave. There is another term we use for the bands that are located between the AM broadcast band and the amateur 10-m band. We call these the *high-frequency bands* (or HF bands, for short). The HF bands are also commonly known as the *shortwave bands.* A few million people throughout the world are dedicated to tuning in shortwave broadcasts from stations all over the world. These shortwave listeners (SWLs) are fine people who thoroughly enjoy their hobby of shortwave listening. I know several who have gone one step further—from listening—that is, to contacting foreign countries by becoming radio amateurs. (A few SWL addresses are listed in the appendix.)

Each of the HF amateur bands (in which we do about 90 percent of our operating) offers its own special brand of operating entertainment. Some are useful for short-range contacts, some for long-range.

For example, someone in New Jersey who is interested in regularly contacting someone in Pennsylvania would more than likely use the 80-m band or the 40-m band because the lower-frequency bands are the shorter-range bands. However, if I wanted to regularly talk from New Jersey to California, I might instead use the 10-m or 15-m bands. Those higher frequency bands are best for the long haul.

Take a minute here to look at how the band numbers, in meters, go opposite to the band numbers in the frequencies; that is, the 80-m band starts at 3.5 MHz, whereas the 10-m band starts at 28.0 MHz. The meters go down while the frequencies go up. Well, I won't bother you with the full technical explanation of this except to say that the meters refer to the *wavelength* of a radio wave; that is, the distance one radio wave travels. We call that its wavelength, which is commonly measured in meters (instead of feet or yards). While it isn't important enough to worry about, there is an easy way to figure out how many meters there are to a given frequency: All you do is to divide the frequency in megahertz into 300 to get meters; or, divide meters into 300 to get the frequency in MHz. Just to give you a quick example, 10 m divided into 300 gives 30 MHz. Well, the amateur band isn't really at 30 MHz, as you can see from the chart (it is 28.0 to 29.7 MHz), but that's close enough for us—we're just looking for a sort of nickname, so we call it 10 m.

It's that way for the other bands too. If you wanted to be exact, the 80-m band runs from 85.71 m (300 divided by 3.5 MHz = 85.71) to 75 m (300 divided by 4 MHz = 75). It would be awkward to call this the 75- to 85.71-m band, wouldn't it? We happily go on our way calling it the 80-m band. It's much simpler that way, and we all know what we are talking about, as will you when you begin your novice studies.

Let me tell you a bit more about the characteristics of some of our HF bands, after which I'll get into the frequencies of those bands in which novice operation is permitted.

Take the 10-m band. Most of the time this band is dead at night and becomes useless. But during the day it can be a very exciting band. It takes very little transmitter power to cross the country on 10 m due to something called *skip*. Skip is what happens when radio waves bounce off the ionosphere (upper atmosphere) and come back to earth great distances away. Ten meters is a band that all novices enjoy operating. Your *CQ broadcast* (call to any station) can just as easily bring an answer from someone across the country as it can from someone in Europe—and it will come in as strongly as it would from right down the street. You don't need high power to have fun on the 10-m band. We refer to the 10-m band (and also the 15-m band, to some extent) as a *daylight band* because it is mostly open only during the day.

Now, the 20-m band is the DX (foreign country contacts) band. It is open 24 hours. As you can guess, it is our busiest band. Unfortunately for you, novices can't operate there. No problem. When you get active and learn what 20 m has to offer, you will also gain the incentive to upgrade to the next higher class of license so that you can operate on 20 m.

The 30-m and 40-m bands are also 24-hour bands. They don't carry out as far as does the 20-m band, but they offer their own special brand of fun. For one thing, they are less used, and chances for a contact are better because you don't have to put up with as much interference from other amateur stations.

The 80-m band, opposite to the 10-m band, is a solid, reliable band for short distances. It is mostly closed during the day, until late afternoon. It stays open right through the night and early morning hours, often permitting much DX, particularly on cold winter mornings, as I've written about in another chapter.

In all the variations to these bands, each with their own special uses, my favorite and the one on which I have always had the most fun is the 40-m band. That's probably a reflection of the fact that I like to gab over the air, and for that purpose, the 40-m band is one of the most solid and reliable bands we have. We all develop our favorites depending on our interests. We make use of all bands too. Modern radio sets have all of the HF bands built into them, which lets us switch from band to band.

Within each amateur radio band, certain frequencies are assigned for operating voice, and certain frequencies for operating code. Further, as you can see by the areas of the diagram, to the right of our apartment building, there are four bands permitted for novice operation: the 80-m band, 40-m band, 15-m band, and the 10-m band.

Within each of these novice bands are the assigned frequencies for novice operation. These are:

80 m: 3.70 to 3.75 MHz
40 m: 7.10 to 7.15 MHz
15 m: 21.1 to 21.2 MHz
10 m: 28.1 to 28.2 MHz

To give you an idea of how much room there is to roam around the novice bands, the 10-m novice band ranges from 28.2 down to 28.1 MHz; it is 0.1 MHz wide (28.2–28.1). Do you remember from earlier in the chapter how to convert from MHz to kHz? (You will have a question on it in your novice exam.) Just to refresh your memory, 1 MHz is the same as 1000 kHz. So, then, 0.1 MHz is the same as 100 kHz. Or, figure it all out the long way.

$$28.2 \text{ MHz} = 28,200,000 \text{ Hz}$$

$$28.1 \text{ MHz} = 28,100,000 \text{ Hz}$$

$$
\begin{array}{r}
28,200,000 \text{ Hz} \\
- 28,100,000 \text{ Hz} \\
\hline
100,000 \text{ Hz} = 100 \text{ kHz}
\end{array}
$$

Anyone can operate novice frequencies—in fact, I often get on the novice frequencies to look for and welcome a new member to the hobby—but novices can operate only on novice frequencies. This is as it should be; it gives you a chance to meet other more experienced operators who can help you to develop your operating skills.

That's about all there is for you to know about the amateur bands for now, since you have already learned what you need for the novice exam. The information from this chapter will stay with you throughout your novice studies. You now know where we operate; that is, you know where the amateur radio bands are in the radio spectrum. You will be entering novice studies with a far better understanding of amateur radio than have most other novice students.

# · 8 ·

# Call Signs

Every amateur radio operator in the world, no matter the country, has his or her own personal call sign identification which they use on the air. My call sign is K2VJ. Call signs are a combination of letters and a number. Every country has its own special number-letter combinations so that operators can tell which country they are listening to just by the call sign.

Amateur radio operators in the United States have call signs that start with one or two letters, followed by a number from zero to nine, and ending with one to three letters after the number.

About that call sign number: The United States is divided into 10 call sign districts numbered zero through nine, plus Alaska and Hawaii. Call sign districts start in the New England area with number 1; the second district is New York and New Jersey; the third district is Pennsylvania, Maryland, Delaware, and Washington, D.C.; the fourth district is the Southeast; the fifth is the Southwest; the sixth is California only (always the most active ham radio state in the country); the seventh district is the Northwest; the eighth is the Mideast; the ninth is the mid-Central; and the zero (or tenth) district is the Midwest.

In the early days of amateur radio, letters were not included before the numbers; instead, they were used only after the number. For example, W1AW was 1AW. It wasn't until the 1930s that the Federal Communications Commission (FCC) set up call signs with letters before the number. All call signs started with a W, back then.

However, following the large increases the hobby experienced after World War II, the FCC began issuing call signs with a prefix (first letter) of K.

For example, my original call from Pennsylvania, issued immediately after World War II, was W3MLZ—third call sign district, when W prefixes were still being issued. Later, in moving to Oklahoma in the 1950s, my call sign became K5QHL— fifth call sign district, and no longer with W prefixes available.

Then the K prefixes ran out, so they began to issue two-letter prefixes, starting with WA. My call sign when I moved to New Jersey in 1960 was WA2TVG.

The FCC has changed things quite a bit since then. They now issue U.S. call signs with prefixes that start with letters such as AB, KA, KB, KR, KO, WR, ND, AL, WH, in fact, just about anything that starts with A, K, N, or W.

The variety of prefixes now used in the United States provides enough reserve call signs to last into the next century, unless this book converts a few million lucky Americans into the super hobby of amateur radio.

There was also a time in our history when moving to a new call sign district meant we had to apply for a new call sign. But that is no longer necessary either. The FCC lets us keep our old call sign if we want, and many of us do because we get attached to them. In a noisy crowd, we can hear someone whisper our call sign before we can hear someone shout our names. Radio amateurs are like that.

Radio amateurs who keep their old calls when they move sometimes create some rather interesting situations. For example, when someone from the sixth district (California) answers our CQ (call to any station), we sometimes find that they're right down the street from us, and not in California. They didn't change their call sign after moving from California.

My present call sign K2VJ is from my first and middle initials, VJ. And, no, I didn't get this special call sign from friends in high places. Not at all. I got it at a time when the FCC was more generous in permitting those of us with the top level of license to select our own call from the list of available call signs. K2VJ was available so I grabbed it.

I'm glad I did. I've gotten a lot of mileage from it, and the call is recognized in a few amateur radio circles, these days. If I ever did move to sunny California or anywhere else in the United States, I would hold on to this call sign because it's a personal thing with me by now. We radio amateurs are like that.

# · 9 ·

# Contests and QSL Cards

Contests and QSLing are where we get into a beehive of activity in amateur radio, that's for sure. You might not think so, and you might wonder what's so good about it, but I can tell you that just about all of us get into contests, one way or another, one time or another.

QSLs, which I'll explain shortly, are postcards we send to each other to acknowledge ("QSL," in amateur radio talk) a contact on the air.

Yes, if there is any one activity we have that overshadows all others in amateur radio, it is our forever competitive nature that leaves us in a nearly helpless condition when the lure of contests comes up. But there are so many amateur radio contests, where should we start in describing them?

Probably the all-time favorite, and our grandaddy contest of them all, is that of Working All States—"WAS," as we call it. Nearly all newcomers to the hobby gear up for this award.

Getting WAS is simple. All you do is contact at least one station from each of the 50 states.

Well, maybe it isn't so simple after all. The usual novice

situation is one of limited equipment and limited resources, and newness to the hobby that hasn't allowed the novice to yet learn how to go after the far-out states—but this all adds to the fun of trying.

When going for the WAS award (which, like all other ARRL awards, gets you a very attractive $8 \times 10$ certificate with your name on it), you will find that if you live in the western part of the country, you will become absolutely convinced that there are no radio amateurs in states like Vermont when you need Vermont for your fiftieth and last state. And if you live in the eastern part of the country, Wyoming will seem to have no radio amateurs. They do, however, and you will eventually find one after a bit of work but a lot of fun.

My son, Jim, had to work hard to find an Alaskan station. You would have thought he was searching for an outer galactic signal, considering the problems he had. He was using an old, heavy-as-a-boat-anchor radio set at the time, and he burned a lot of midnight oil until he finally struck paydirt. As far as he was concerned, that Alaskan station might just as well have been from Alpha Centauri.

Ask any newer radio amateur, "What was your fiftieth state?" You'll get a positive response because they all remember: They worked for it and had fun at it.

It is even more fun when you discover that making the contact is not the whole story. You need proof that you made that contact. The American Radio Relay League (ARRL) issues the WAS certificate, and they want it in writing that you made those 50 contacts. How do we do this? We swap QSL cards to confirm a contact.

QSL cards are postcards which have our call sign, address, signal report, date, time, band, equipment, and whatever else we like to put on them. (See Jim's QSL card on the dedication page to this book for an example.)

Only after you receive a QSL card can you prove you made the contact. Not until then does the contact become official. You can see why QSL cards get to be pretty important to some of us. (It is not required that QSL cards be sent with each and every contact—we send them only when requested or when we want to request one from the other station.)

Some QSL cards are quite fancy, the extent of detail being up to the owner's tastes and budget. Most of them can be obtained from the many printers who advertise this work in the various amateur radio magazines (these magazines are described in Chap. 22). They aren't too expensive. Some radio amateurs go all out, as did one of the members of my chess and amateur radio club (CARI, described in Chap. 42) who had an artist put exotic chess pieces all over the card. Mine is a standard size (3 × 5 in) on which I was "humble" enough to put my picture. (I figure we talk to radio friends for years and never get to meet them or have any idea what they look like. I wish everyone would go the photo-QSL route.)

Jim's QSL card also includes a photo, as you have seen. The picture was taken when he was 12. Now that he's a 6-ft college kid, and twice the size, the rascal still uses those cards (he has a zillion of them left over!) because, I suspect, when he sends them out to an obscure country, he figures they'll give a kid a break and return QSL cards more readily. Maybe it works. This business of getting a return QSL card from foreign radio operators is a big part of the action. They sometimes need to be coaxed to send a card, which is what Jim might have done.

QSL cards from foreign countries which contact large numbers of U.S. amateurs are sent in one package through international QSL bureaus. Or the foreign operator may have a QSL manager in the United States who mails his cards to American operators when receiving the foreign operator's log sheet with those contacts listed on them. Over 2000 U.S. radio amateurs serve as QSL managers for other radio amateurs around the world.

By the way, if you are also interested in stamp collecting, particularly stamps from other countries, being a QSL manager is a fascinating sideline to the hobby which offers both fun and profit. Yes, they make a few dollars from it, and it is a perfectly acceptable and honorable way to acquire extra income. I won't go into it here, as it gets rather complicated, but if you ever become interested in the idea, just continue with your reading because the path to becoming a good QSL manager is to become a good DX'er (contacting foreign stations). By the time you acquire a few DX awards, you will have learned exactly what you need to know in becoming a QSL manager for a DX station from some remote country. (DX and DXing will be described in chaps. 10 and 11.)

How big an activity is QSLing? The American Radio Relay League's overseas QSL bureau ships about 2 million cards to American amateurs each year. The unofficial estimate is that maybe 10 million cards are being shipped around the world every year. There was a time in Jim's DX work when I thought that about half those cards were being sent to our house. His cards are now hanging on all the walls of his bedroom, and the rest are stuffed into several drawers in his desk. When visitors come, they just stand there in awe scanning up and down those walls. And when I whip out huge stacks of more cards from foreign countries, we usually get someone else interested in amateur radio.

As I said, contests are the backbone of amateur radio activities, and they are all up to your own preference. We seem to have contests of one sort or another every week, but that is an exaggerated condition imagined by those who don't go for them very much. Unless one's interests run toward a devoted pursuit of contests, one's cup can surely run a wee bit over the brim. Still, they are among the finest activities by which to brace a cold and lonely weekend.

New contests come along every day in amateur radio. There's no way you can keep track of them all, although most contest sponsors get them publicized in one or another of the amateur publications.

A recent issue of *QST* (publication of the ARRL) had 46 different contests listed for the month. One day, I chanced across one of them, the Foxboro Co. Amateur Radio Club of Bellingham, Massachusetts. Here's how this particular contest went: Stations were to contact three of their club members who were operating on the same band. Each club member would hand out a letter spelling F-O-X, along with a number. The idea was to contact three club station members on the band. I took the time to search for the other two, after having contacted the first by accident, because the idea intrigued me. Later the club sent me a very attractive 8 × 10 certificate. You could collect dozens of these certificates every month if you were interested.

And contests are not exclusively an American pursuit, even though we Americans are best known for chancing upon any situation under any circumstances, and if it flies, runs, or crawls, we'll put a bet on it. One such contest consists of contacting each of the 3077 counties in the United States. Who do you think considers this to be one of the greatest contests around? Amateur radio operators in Japan, that's who. And they get mighty serious about it too.

# · 10 ·
# DX

Without a doubt, the "royalty" of amateur radio contesting is the noble pursuit of DX—contacting foreign countries.

On a general level, DX is simply an abbreviation for the word distance. But in our amateur radio work, DX has a few different meanings, depending on the nature of a conversation.

For example, on the radio bands where the normal range of contact is, say, 100 mi (miles) DX could mean someone in a state a few hundred miles away. We then say we "worked DX" if we made that unusually distant contact on a local band.

But the big meaning for DX is contacting foreign countries. And if it is someone new to DXing who is talking about their contacts, DX might refer to countries in Europe or South America. But when an expert refers to DX, you know he or she is talking about an obscure place such as Bouvet, Qatar, or Franz Josef Land—wherever they are. See what I mean?

We look for awards in our DXing, awards which serve as milestones in the numbers of countries we contact and as a boost to our keeping up with the other DXers.

The first level of awards in DXing is the glittering DXCC award—DX Century Club, which we enter from having contacted (and received QSL cards from, don't ever forget!) 100 different countries.

Now, this might sound like a difficult chore, but it really isn't all that hard. The first 40 or 50 contacts are made easily enough, while the next 20 or 30 get to be a bit more work. It's those last few that keep you glued to the set when you should be sleeping or out in the fresh air. There is always something— call it Murphy's law—that says if you need just one more country, even if you haven't yet made use of half the easy ones in Europe, it will still seem like weeks before you reach the hundredth country.

You will have plenty of company on the airwaves in looking for DXCC. Over 30,000 hams, worldwide, have qualified for this certificate since 1945. The DXCC office at ARRL processes tons of DX cards every month, by and for those who are after DXCC. The ARRL says DX cards increase by about 30 percent each year, which gives you an idea of how this fascinating search for DX affects so many of us.

Just why is DX such a super activity? I would try to answer, but you wouldn't believe what I said. You see, there is an inner excitement to DX that comes not from reading about it but by doing it. DX is one activity you must experience to appreciate; it is something of a bug, a virus, and there just isn't any cure for it.

One of the problems is that when you reach DXCC, you find yourself on a runaway locomotive from which you may not be able to get off, to quit and rest on your accomplishments. Some could, but others who have been bitten by the DX bug seem so caught up in it that they go on to the next level of award, DX2CC—contacting 200 countries.

Where does it all end? How do you get off the freight train? One way is to become the world's leader in DX, which is not likely because today's leaders are working on DX4CC, and if you look around, there just aren't 400 countries kicking around Mother Earth. Ah, we've got ourselves a problem now.

How come the leader is so high up the ladder? Well, the one written about later in this book had been at it quite a long time, from back when countries you may never have heard of were around but are now defunct.

We change, you know. For example, amateur radio lost a prefix country when we ended the American presence at the Panama Canal. When we signed off on the "big ditch," we also signed off forever on an official radio-prefix country for U.S. operators who operated there with special call signs—KZ5's—which none of you newcomers will ever be able to work.

On the other hand, a new entry has popped up—China. China had gone QRT (stopped transmitting) several years ago when they banned amateur radio. Now things are different and Chinese operators are back on the air again.

That's the way it goes. The list of countries is forever changing. (Can you see why DX operators get to be so good at geography?) Those who made contact with countries no longer in existence have an edge over the newer DX hounds. But that doesn't stop us from digging in.

You don't need high power to go after DX. In fact, there is a clan of dedicated radio amateurs who take much pride in deliberately using very low power in their DX work, for which they receive special recognition. Good equipment, particularly with antennas, is part of their scheme, but patience and practicing a lot of operating skills are the real needs for low-power DXing. Patience is needed: you had better be naturally equipped with it if you take up DX. Skill you develop as you go along.

Some who have been bitten by the DX bug become professional night stalkers, hunched over their sets late at night when band conditions are best for DX—tuning, listening, waiting, and then pouncing when a new country shows up. And when you add that new one—well, you can read what my son, Jim, WA2JNN, has to say about it in a later chapter. It's a thrill and a huge source of satisfaction that lets you finally go to bed with a smile on your face.

DX hounds join DX clubs which do some fabulous things. DX clubs mail out bulletins to notify their members that some faraway uninhabited island is going to be briefly populated by a dedicated group of radio amateurs who travel halfway around the globe just for the pleasure (?) of lugging radio gear up some steep goat path in an isolated part of the world on what is officially recognized as a "DX-pedition"—all for the purpose of putting a new radio country on the air.

In recent years there have been DX-peditions to such exotic locations as Brunei, Montserrat, Navassa, Papua, and the Islands of Clipperton, Bouvet, Cocos Keeling, Pehrhyn, Peter First, Spratley, and Pitcairn. Frankly, I have no idea where any of these are. But DXers do.

And for these DX-peditions, tens of thousands of radio amateurs, worldwide, lie in wait. A new country! Word goes out quickly, and the bedlam that announces a DX-pedition's first broadcast is so wild that it seems to permanently burn a hole in the band.

DX operators who take part in DX-peditions fulfill a lifetime of operating glory in only a few days. A recent DX-pedition to Navassa (30 mi west of Haiti) made 34,000 contacts in 6 days—that works out to four contacts every minute, 24 hours a day for 6 days. The lives of such DX-pedition operators can never return to "normal," for one cannot operate at an obscure DX station and come away without a changed perspective.

# · 11 ·
# DXing

Radio conditions for DXing (foreign country contacting) are usually best late at night, which is true of our favorite DX band  ✗ of 20 m. Some other bands peak during daylight hours, but they are more flighty and less reliable for pure DX than is the good old 20-m band.

For 20 m, the wee hours are best for two good reasons: Not only is propagation (reception conditions, see glossary) better, but this is also the hour when most other U.S. operators are asleep and less likely to be bothering you with their competition for contact with an obscure country.

There is a 5-hour time difference between the eastern United States and England. Their early morning, right before dawn, is a peak operating time which works out to be about 1 a.m. or so, our time. If you want to work into a rarely contacted country in Asia, for example, and if their period right before dawn is their best operating time, figure what time that works out to be in the United States. You can see why some DX hounds look for jobs with a swing shift so that they can be up at all hours of the night to work DX.

Let's back up to that new word I used—*propagate*—to get

45

a little bit of understanding about it. It won't be a technical explanation, rest assured.

My dictionary defines the word propagate: "Spread, pass on; send further." Okay, none of that is technical, is it? So let us simply apply that definition to radio work because "sending further" is what DX is all about, and when you have an idea of what we are talking about with propagation, you will also have an idea of how we go about DX.

As radio signals leave an antenna, they head out into space. On the bands most commonly used for amateur radio DX (20 m, for example), these radio signals are reflected (skip) back to earth by a sort of blanket in the upper atmosphere which is called the *ionosphere*.

Oh, oh! A new technical term—*ionosphere*. Let's see what my dictionary has to say about ionosphere: "The ionosphere is composed of layers of atmosphere ionized by solar radiation which allows transmission of certain radio waves over long distances on earth by reflection."

Well, that's just about what I said in much simpler words— an upper-atmosphere blanket that bounces radio waves back to earth for DX.

During daylight hours, the ionosphere acts like a mirror to reflect radio waves back down to earth. At night, however, it thins out, and, instead of reflecting radio waves, it causes them to take a curved path in the ionosphere before it reflects them back to earth. This curved path is stretched out at night so that the reflection takes place higher, causing the reflected radio wave to come down farther away. In other words, we get better DX because the reflected path is longer. That is why reception is much better on the shortwave bands at night than it is during the day. End of lesson.

But other things can happen to that ionosphere. We go into what is known as the *sun-spot cycle* (see glossary) every 11 years. In this 11-year cycle, the ionosphere is thicker or thinner, and DX is better or worse. We didn't learn about this effect until right before World War II. You will read a reference to this in a later chapter about when radio amateurs first started using the 10-m band.

Mixed in with the 11-year cycle is the occasional flare of

solar radiation (see glossary) which plays havoc with shortwave reception. You notice this when a TV program you have been watching (if you are on an antenna, and not cable) starts to fade out and suddenly there are TV stations coming in on all VHF channels. This tells radio amateurs to get cracking—DX is hot!

I recall a personal incident of freak propagation when, a few decades back, I was radio officer aboard a ship on which I had an amateur radio station. At the time we were in the Pacific off the coast of Mexico.

One afternoon the 10-m band, which had been completely dead, suddenly shot alive, and I heard stations coming in from every part of the United States. They were booming in, too. Yet no one was talking to any other distant station, only to local stations.

So I called a CQ (general call to any station). When I stood by I was absolutely shocked to hear dozens of stations calling me! To make the story short, it turned out the band was completely dead within the United States but was open only between the entire United States and the location of our ship in the Pacific. I was the only station on the entire band that anyone could hear, DX-wise, and I was able to hear stations in every part of the country. It was as though I had gone to radio heaven; I was suddenly a DX station!

For about 1 wild and wooly hour, I worked one station right after another, about 200 contacts, until the skip conditions abruptly changed and the band again went dead.

While it lasted, I was the only rooster in the hen house, from which I gained a memory I'll never forget. For all radio amateurs, one time or another, special situations provide us with special memories which we take delight in describing. This hobby, more than most, seems to be like that.

Every year, in early Autumn, we welcome ourselves back to a season of operating with a huge bash in the form of a worldwide DX contest. Some stay at it the entire weekend, seldom leaving the radio set to cool. Working 100 countries in this one weekend is possible. Some work all continents in an hour. It gets to be some wild fun.

And to keep us on our toes, there is another giant DX

contest right around the peak of the winter season for which some DXers, who really aren't bug-bitten DXers, join in as a way to get rid of all their DX infection with one "shot." They then go back to living normal amateur radio lives afterward. You can believe that a weekend in a DX contest is a fine test of both operator and machine.

Yet don't lose sight of the fact that DX, like the many other activities we have in amateur radio, is a personal choice. You can take it or leave it. No one says you have to do this or that; it is entirely up to you.

Now, with all the raving I've done about DX work, would you like to guess how many countries I have personally contacted? Do I have DXCC? Answers: (1) Maybe two dozen. (2) No. What's wrong with me that I'm not into DX? Well, I'd never have had the time to write this if I were. I like DX and I get as much a kick from it as anyone, but I'm able to pull the "big switch" and walk away when I don't find a new country.

So far, anyway.

# · 12 ·

# Sending Messages

One of the more interesting activities that radio amateurs voluntarily take part in is sending messages around the country and to many parts of the world. And quite a busy activity this is too.

Message handling goes as far back into amateur radio's history as the hobby itself, so don't skip this chapter thinking it isn't for you. You may have a few surprises in store. You should give a second thought to why many radio amateurs are so dedicated to the fun of sending messages. Fun it is too, or it would never have built up into as large an operation as it is today, nor would it have lasted this long in our history.

Using scheduled times and frequencies, we get together on traffic networks (which we simply call *traffic nets*) to relay messages from others or to send our own messages all over the country and to many parts of the world. (Some countries don't permit this sort of activity by their radio amateurs—these include many European and Asian countries.)

Our messages must be noncommercial, of course. The laws do not permit us to compete with commercial services. We are, after all, a nonprofit hobby.

Message handling is an art which has a way of developing

49

your on-the-air operating skills like no other activity we have. Even with all my past years of handling commercial radiograms, I am still impressed when I eavesdrop on traffic nets to hear their rapid-fire, efficient exchanges.

Then, again, they need to be well organized because, when 30 or 40 stations check into a net, the net control station must keep order or else things fall apart.

There is, indeed, something about traffic handling that displays the art of radio communications at its best. When you get your radio amateur license, just remember this one rule: Better listeners make better talkers. You will catch on.

Yes, there are traffic nets for novices. Everyone in these novice nets (including the advanced operators) sends code at the slowest speed necessary to make sure novices can copy them. No one tries to show up anyone on the novice traffic nets. If you do get interested in these, write to the ARRL for a listing of all traffic nets in the country, and from the listing you will find some nets for novices within your range and at times and frequencies that meet your needs. You might be surprised at how many of these nets are going on and at how many people in the hobby take part in them.

By the way, let me pick up on that idea of becoming better operators, because, among all the certificates that are awarded by the ARRL, there is none so prized as the one known as the A-1 Operator Club certificate. This certificate says on it: "Membership in the A-1 Operator Club represents adherence to the several principles of good operating: (1) Careful keying and good voice operating practice; (2) Correct procedure; (3) Copying ability; (4) Judgment and courtesy." A *QST* item explained: "The more you covet this award, the less likely you are to obtain it. You can't apply for it. You can't buy it. It may take a few years to qualify. You may have already qualified and still not have received this award. Although deserved, it may never come."

Then how in the world do you get it? By constantly practicing good operating and hoping that someday two current members of the club decide to nominate you. And when that happens, you don't have to brag about it except to quietly mention it in a conversation. This is one award that speaks for itself.

Let me now tell you something about our amateur radio National Traffic System (NTS). NTS consists of (1) local traffic nets (from nearby areas); (2) section nets (sections of states); (3) regional nets (which are the same as the amateur call sign regions); and (4) area nets (which are divided into three zones corresponding to U.S. time zones).

Suppose my neighbor were to give me a call one evening to send a radiogram to a friend in California. I would check into my local net, the Vineland, New Jersey, 2-m repeater net, at 10:30 p.m. The following morning, someone would relay that message to the New Jersey section net at 10:00 a.m. It is then picked up by someone who relays it to the second district regional net (consisting of New Jersey and New York) at 1:45 p.m., from which it is relayed to a representative of the eastern area net at 2:30 p.m.

On transcontinental messages, the area nets have members from what is known as the TransContinental Corps, TCC, who meet with their counterparts in private schedules on a one-on-one basis; this takes place eight times per day for each of the three time-zone areas; 7 days per week.

Once my California message is picked up by a TCC rep, it filters back down through the system to the city in California for which it is addressed. If the address I gave included a telephone number of the addressee, it could very well be phoned in for delivery to that person the day after I sent it out from New Jersey.

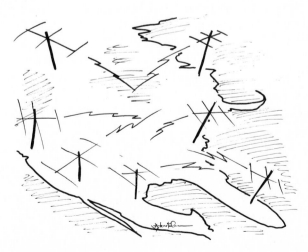

For messages going outside the country, say, to Australia which permits amateur radiograms, if I had gotten the message to the New Jersey section net by 7:00 p.m., I could have given it to a representative from the second region who would take it to a 7:45 p.m. net from where it would go to the eastern area at 8:30 p.m. The eastern area net also has a rep from what is known as the International Assistance Traffic Net, IATN, who could be from some other country or who could be a U.S. amateur with ready access to Australia on a daily basis. The IATN rep works one-on-one with a counterpart in Australia every day.

We have 3 area nets, 10 regional nets, about 100 section nets, and several hundred local nets, all meeting daily, and some, several times each day.

Who is allowed to send messages by amateur radiogram? Anyone and everyone, as long as the message is not intended for commercial purposes such as advertising or quoting prices for store items. (Note that we can send information on personal equipment we have for sale if we are not in the business of selling such equipment.)

What kinds of messages can we send then? Just about anything that is not commercial in nature. We send greetings, for which there are standard codes. That is, if you wanted to send the message, "Greetings on your birthday and best wishes for many more to come," we wouldn't have to send anything more than ARL46, and the delivering station would read out the full greeting. But in the process mentioned earlier, all those stations who relay this message would send only the five characters ARL46. You really ought to get to know a local radio amateur because we can do a lot for you in keeping you regularly in touch with old friends around the country, and at no cost to you.

These traffic nets are business-like while they are in session, but I have often taken part in many which close and then go into friendly conversations. In fact, many of these traffic people are old friends, and each check-in station is given an opportunity to make comments to others in the net. These after-net conversations get to be something of a family affair at times, although there is always a point in the procedures when they take the time to welcome newcomers. Join in when you are licensed. It doesn't take long at all to become "one of the gang."

One of the most enjoyable features of amateur radio forms is the message-handling spin-off known as the *phone patch*. Phone patches are devices which connect our radio sets to the telephone lines. Yes, they are legal.

I used to do a lot of phone patching to my daughter, Barbara, in Raleigh, North Carolina. My amateur friend there would weekly meet me on the air, and he would connect his phone patch in for me to talk to Barbara. She got to be rather expert at talking over the air via the phone patch. Our conversations would carry on just like an ordinary phone call. In fact, we have *voice-operated* equipment which permits us to switch from transmit to receive automatically just by talking into the microphone.

Sound simple? It is basically. Of course, it doesn't have all the convenience of a direct phone call. Also, no one has exclusive rights to any amateur frequency so we sometimes have to talk through interference. Mostly, however, good radio amateurs give priority to a phone patch and stay clear.

And our friends at the long-distance telephone companies are not neglected in all this. We still use the telephone for direct calls, and quite often too. I suspect our use of phone patches sometimes ends up generating more direct phone calls than we might otherwise have made because there may be something we want to continue to discuss in private or there may be an unfinished matter to discuss that started on the phone patch. My phone bills prove all this to be true.

(You will later read of the fantastic phone-patching record of Senator Barry Goldwater's station, K7UGA, in letting U.S. overseas troops talk to family and friends back home; in fact, a few hundred thousand such calls have been made over several years.)

My involvement in traffic handling ended up as something big in my life. I used to play chess by radiogram with my nephew, Dave. He was not a radio amateur himself, but he would contact a local amateur who would relay his chess moves to me. I would check directly into one of the eastern Pennsylvania section nets to relay my messages directly to Dave via the station local to him. More than one net member, obviously not chess players, would be startled by such messages as, "King's Knight takes Queen's pawn."

The offshoot of all this chess radiogram handling was the organization my son Jim, WA2JNN, and I started called "Chess & Amateur Radio International," which you will read about in a later chapter.

Are you beginning to see what I mean about this hobby of amateur radio having something for everyone?

# · 13 ·

# Emergency Preparedness

> We were tracking so many funnels that day . . . we were scared. . . .

April 10, 1979, disaster struck Wichita Falls, Texas, when the first of several tornadoes touched down in the city and cut a mile-wide path of destruction that stretched out for 30 mi. Within minutes of the strikes, 20,000 people were suddenly homeless and the entire city was without power or telephones.

What has this to do with amateur radio?

Included in the amateur radio creed is the need for our participation in emergency preparedness. But let me make one point perfectly clear; we are only encouraged to be ready for emergencies as a personal choice—there is nothing in the regulations which says we must do this or that. It is all up to each of us.

At Wichita Falls, only 1 week before the strikes, local radio amateurs had conducted a practice disaster drill. Little did they know.

Within an hour after disaster struck Wichita Falls, local

Amateur Radio Emergency Service (ARES) members had radio communications links set up for the Disaster Service of the Red Cross. Twenty amateur mobile units were scattered throughout the area to serve as spotters, reporting directly to the National Weather Service (NWS). NWS could track tornado cells by radar, but they couldn't tell when or where one had touched down—only personal observation could, which was the service provided by dedicated radio amateurs who risked their lives in public service.

When it was over, the area looked like a war zone. Officials credited the relatively low loss of life to the fact that ARES members did their "thing" in on-the-spot reporting.

Do you remember Three Mile Island, near Harrisburg, Pennsylvania? Radio amateurs had set up a communications link for that occasion between the Pennsylvania Emergency Management Agency, the Office of Emergency Preparedness, the Red Cross, and a local radio-TV station.

True, TMI was a different situation. Usually the disaster happens and then there is the call to radio amateurs for help. Not this time, however. This time the call for help came before the disaster—the disaster that, most fortunately, never happened. But we were there.

Also, until that time, all planning in preparedness for a nuclear situation had been on an attacking enemy. No one gave a thought to the possibility that the "enemy" might be ourselves.

Some of these incidents put the emergency preparedness role of amateur radio in the public spotlight, but many do not. For example, did you know that radio amateurs throughout the country were mobilized and tied into national systems when Skylab threatened to do a number on us, years back, when it fell down from space? We were there—at police stations, city halls, emergency management posts, everywhere—thousands of us.

Preparedness is a big part of amateur radio. It has always paid a dividend to the country, a fact never better proven than when 25,000 self-trained radio amateurs provided Uncle Sam with a ready pool of experts at the outbreak of World War II.

Emergency preparedness is not all drill and work. Nearly

50 years ago, ARRL established the first Field Day contest, which is a weekend camp-out for the tens of thousands of radio amateurs who head for the woods each June to set up and operate on emergency power, attempting to contact as many other stations as possible that weekend.

Rules for Field Day say that points are given not only for the number of contacts made but also for the kinds of equipment used. Scores are sent to the ARRL for establishing winning stations. The entire activity, which runs from early Saturday through Sunday afternoon, gets to be a fun affair for which some wait all year.

And fun it is. Even though most clubs are like ours in having held Field Day outings for years and years and in having the equipment down to where we do the same thing year after year, a year still doesn't go by that Murphy's law (see glossary) doesn't strike in the form of bad connectors (after the beam is hoisted up a 40-ft tower), generators that won't generate, reliable transmitters that won't transmit—it's all part of the game.

But that's what Field Day is all about—fun with a serious purpose; a full-scale shakedown of our ability to set up an emergency preparedness station when called upon; a shakedown of both equipment and people.

We radio amateurs don't merely say we are ready for trouble: we try to make sure we are. Field Day helps us get ready. And radio amateurs all over the country have been proving our emergency preparedness nature by our frequent presence at emergency situations.

# · 14 ·

# Women in Amateur Radio

Back in 1901, when Marconi heard the first radio signal sent across the Atlantic, amateur radio soon began to firm up its place in the new field of radio communications.

They didn't call it "radio" in those days; it was called "wireless," which is sort of like the way they called the car the "horseless carriage" because it moved on its own, without the horses that all carriages needed up to then. With radio, wireless meant talking without the wires that telephone and telegraph needed at that time.

The men of those times who practiced wireless communications were, in fact, state-of-the-art experimenters, particularly those who fell into the class of amateurs, as did so many of the time.

But did I say "men"? Well, pardon me. Let me hurry to assure women readers that amateur radio was never an exclusive field for men. Not at all!

Records show that back in 1915, Emma Chandler, who was later licensed as radio amateur station 8NH, actively operated her wireless station from the town of Saint Mary's, Ohio.

Nor was Emma the first YL (YL, an accepted term in the hobby that we've been using from its very beginning, means "young lady"). Unfortunately for history, there is no record of who (other than Emma) might have been our very first YL in amateur radio.

But it is a fact that amateur radio has always been blessed by the active participation of women in the hobby. Their accomplishments, organizations, activities, and dedication to the hobby can be seen from their numbers—coming up on 30,000 in the United States alone, and a few thousand other women radio amateurs in countries throughout the world.

YLs have formed the Young Ladies Relay League, an organization that is similar to the American Radio Relay League.

YLRL was formed in 1939, and it publishes its own journal. Each amateur radio district elects officers to YLRL. Their annual conventions draw representatives from every state and from several countries on various continents.

YLRL also sponsors its own operating awards such as the WAS/YL for contacting a YL in each of the states; WAC/YL for contacting a YL from each continent; and there is the challenging YLCC award given for contacting a YL from each of 100 countries. YLCC comes a bit harder than DXCC since it requires contacting a YL, and not just anyone, from these countries. But it has been done by many who tried harder.

There is very little in this hobby of amateur radio at which the YLs have not matched the OMs. (OM is an accepted term in the hobby that's also been around forever. It means "old man," but in a respectful way. The Y in YL has been our tribute to women who have made themselves special, in our estimations, by becoming amateur radio operators.)

There are no amateur radio operating activities at which women do not match the OMs: traffic handling (for which the all-time volume record goes to a YL); DX; setting up YL Field Day stations; attending and participating in club meetings; and more. I don't speak for all OM members, of course, but I do speak from my years of observation when I say that the presence of a YL at a meeting or a traffic net or even a DX pileup (that is, when dozens or hundreds of stations greet a new DX station on the air, everyone trying to make contact with the DX) is, in a way, special. But on the air everyone is equal, particularly in operating by Morse Code. On the air, we are judged by our operating talents alone. In a later chapter you will read the comments of a few YLs I have quoted. What they have to say is said in their own words and in ways I could not match.

I do, however, want to tell you here about Cheryl Maple and Mike Epperson who met while they were students in a novice course. They soon received two sets of licenses; their ham licenses (WB0ZAR and WB0ZAN, respectively) and a marriage license. What made this incident even more special was the fact that the other students in the same novice course were Cheryl's mother, Ursula, who became WD0FPR; dad, Stan, who became WB0ZAQ; sister, Maxine, who became WD0FPQ; and Uncle Laveen, who became WB0ZAX. Talk about team work!

There are several other family teams in amateur radio. And we also have quite a large population of husband-wife teams—so large, they are now nationally organized in their own club. Yes, there is something special about amateur radio— something you never read about in technical books on the hobby.

# · 15 ·

# Big People, Little People

Donny Osmond, Ronnie Milsap, Chet Atkins, Senator Barry Goldwater, Arthur Godfrey, General Curtis LeMay, Larry Ferrari—these people have something in common you might like to know about. They are licensed radio amateurs.

Now, if they, with their busy schedules, can take the time to learn about the hobby, become licensed, and get on the air, perhaps you should consider following suit. They each had their own reasons for joining, reasons I'd guess to be about the same as most others give. You will read the personal stories of some of these in later chapters; stories as they have given them to me by phone, tape recordings, or letters.

It is about time in this book that I bring up a term that is new for you, but one that is among the oldest in amateur radio—*hams*. The people I just mentioned are hams, the amateur radio kind of hams. No one is sure how radio amateurs got to be called hams; one version says it came from the Cockney pronunciation of *amateur* as "h'amateur" and shortened to "h'am," or "ham." Whatever, we are popularly known as hams, and most of us have no problem with that name.

Is ham radio the ulcer cure that some claim it to be? Well, it sure does wonders for some in separating the day's hassles from the night's living. Some evenings, we come home from work tired, and we take to the ham radio for a change of pace. Usually the ham station (we call it "ham shack") is set off in an out-of-the-way spot in the house, like the basement.

We tune in, perhaps with the earphones on to further shut us out from what's going on around us, and soon we are talking to someone in Tacoma or Tokyo or Tanganyika. Can you picture this? You can't beat it for a great way to shift gears and to relax.

This is also a good time for me to give you an idea of how well this hobby is suited to shut-ins. Ham radio brings the outside world to them for a very fine improvement in their daily lives.

If you have a handicapped relative or friend, don't prejudge either them or amateur radio by assuming they may be too small or amateur radio too big, and that the two can't get together. Let them decide, not you. They must decide for themselves because ham radio can do some mighty wonderful things for them.

Well, I started out to talk about the big people in amateur radio, and I did that by mentioning a few of the more famous names in the hobby. But I also added "little people" to the title, and I want to be the first to say that our handicapped radio amateurs are as far from being "little people" as you can

get. Later you will read some of the unbelievable things they do in this hobby, such as how the blind can "see" and the deaf can "hear."

If this book accomplishes no more than getting one handicapped person involved in amateur radio, that single success will be my full reward.

And as for ham radio being a cure for ulcers, it sure can't hurt.

# · 16 ·

# Repeaters

It is true that some radio amateurs consider DX (foreign country contacts) to be the most exciting feature to the hobby. But it is more true that many of us consider the most popular feature of amateur radio to be our *repeater* system.

Repeaters are remote receiver-transmitter stations which automatically receive and repeat what they receive. They receive on one frequency and rebroadcast—repeat—what they receive on another frequency. Thus we can talk to each other through a repeater instead of through direct contact.

The main advantage to repeater use is in the fact that they are always located on some high point—a tall building, a mountaintop, or the like. Of course, the higher the antenna, the greater the distance coverage, which is true of any radio transmission. Some clever radio amateurs have located their repeaters on mountain peaks so high and so remote that they are operated by solar (sunlight) power simply because they are too far from commercial power.

We have thousands of these amateur radio repeaters scattered around the country. Most of them are on the 2-m band (144 MHz). Some are on other bands too, and I'll get into that in a while.

Because repeaters operate on very high frequencies (VHF) where the wavelength is short, their antennas are very short—smaller than VHF TV antennas. Most important to us, however, is the fact that we can contact repeaters with little walkie-talkies using stubby antennas only 3 or 4 in long.

In amateur radio, we call our walkie-talkies *handy talkies* or, for short, *HTs*. When we talk into an HT within range of any repeater, it repeats our call over its entire coverage range. And therein lies the secret pleasures to our amateur radio repeater system—that even with a little HT we can cover distances from 20 to 100 mi, depending on the repeater's location. This means there will always be a few or a few hundred hams listening to each repeater almost all the time. Work on that awhile, to let the obvious advantages sink in.

I'm saying that with shirt-pocket radios you can reach most hams within dozens of miles. And that's not all: We can even reach around the world with them, as I'll soon explain.

The 2-m band sends radio waves out in *line-of-sight*, which means that if we were to try to talk directly between two HTs, the best we might get is a distance coverage of 5 or 10 mi. But with repeater stations located in high spots, the line-of-sight coverage gets to be as far as 100 mi for some of the super repeaters located on top of mountain peaks.

On the other side of the repeater link your voice, as relayed through the repeater, goes through that same high repeater antenna which broadcasts your HT—repeats it—just as far, or farther, than it can hear you. Obviously, having a shirt-pocket ham radio set does wonders for these contacts.

That's not all. Most repeaters have what we call *telephone access*: that is, they are connected directly to a telephone line. If we know the code for a local repeater station, we can have the repeater automatically transfer over to the telephone line for a direct call to anywhere. Most HTs have *touch-tone pads*, which are the same type of push-button dialing as is used on the telephones. Think about that feature for a while.

Imagine you are out for a long walk in the woods, perhaps hunting, when you find you've gone too far in too many directions; in other words, you are lost. Not to worry. One call on the HT and someone will be able to help you.

Or you are out bicycling when you get a flat tire on some remote road. No problem. One call home or to anyone on the repeater will bring help.

Or you are at a baseball game which has gone into extra innings but you know dinnertime is soon coming up. A short call on the HT, right from the stands, will let the family know you will be late.

I was driving on the Pennsylvania Turnpike one afternoon when it got to be time for my regular schedule on the 40-m phone band with my good friend in Massachusetts, John Dould, N1BHL. (John and I have been talking on 40 m almost every weekend ever since he became the first one to join my chess amateur radio club, which is described later in Chap. 42.)

No, I didn't talk directly to John through a repeater. I used a new twist; I contacted a radio amateur by repeater—Bill Barrett, KC3ED, in Shamokin, Pennsylvania—who connected me into his 40-m radio. The hookup went from me, in my car, using only the HT connected to a magnetic-base antenna sitting on the car roof, through the Pennsylvania repeater, to Bill, who plugged me into his 40-m radio. John was listening on 40 m at the time, so we had a fine QSO, a few hundred miles apart, while I was tooling down the turnpike. And all I had was that little shirt-pocket HT. Fun? You bet. But that's not all.

From among the thousands of 2-m repeater stations in the country, several are connected to second repeaters which operate on the 10-m band. You may remember from an earlier chapter that the 10-m band is terrific for great distance coverage with very low power. Well, if you should be walking down the street with an HT in certain areas of the country, you can *access* (contact) a local 2-m repeater which, in turn, automatically accesses a 10-m repeater. To make an exciting story short, several of us have enjoyed talking to England and anywhere else that might be listening to the 10-m repeater, while using nothing more than a little shirt-pocket HT on 2 m. That's fun too, I can tell you.

Some of these repeater systems get to be rather sophisticated. For example, the *Metroplex* repeater system in New York City automatically operates on four bands, including

2 m. It provides 24-hour emergency communications links, has autopatch, and is completely computerized to include capability for long-distance phone use.

This use of repeaters is, of course, limited to noncommercial conversations. You can't use it for anything that involves selling by commercial vendors. We can use it for selling our own gear because that is a private matter as long as we are not in the business of selling radio equipment.

To give an example of what is commercial and what is not, say you are playing a game of softball at the park and there suddenly develops a shortage of liquid refreshments. This being a hot day, you might like to restock the supply. Well, it would be okay for you to call home for someone to bring a few six-packs over to the park, or to ask someone to pick up some to bring over. It would be strictly illegal, however, for you to use the HT to dial up the phone patch for a call directly to your favorite grogge shoppe for delivery of those six-packs. There's a difference between the two situations (although, if you were dying of thirst, anything goes in an emergency!).

Repeater stations are established by and for radio amateurs. Most are owned by club stations, but some are owned by individuals. We have directories of these repeaters. With about 10,000 repeaters active in the United States, the directory keeps growing. Also, many other countries have amateur radio repeaters, which is a fine point to keep in mind should you be traveling in a foreign country—there's no better way to meet and make a new friend in a strange place than through an amateur radio repeater contact.

One member of our local amateur radio club delivers mail for the U.S. Postal Service. He walks around with an amateur radio HT clipped to his belt. He says it's a great way to keep in touch. Other members who have boats pack the HT with them every time they go fishing. With a range of 20 to 40 mi, they know someone is always listening should anything go wrong, or if they just want to make idle conversation when the fish aren't biting.

I have had several fabulous experiences with HTs and repeaters, but the premier of them all is a story that packs a powerful memory I'd like to tell you about, after which you can judge how keen I am on repeaters.

My family and I visited Disney World's Epcot in 1983.

While there, I had my HT along (of course, we take them everywhere we go) and had several conversations with hams in the Orlando/Saint Cloud area. One day I mentioned over the air that my life's ambition was to eat oranges freshly picked from a tree, as I do admire fresh oranges.

Just that quickly, a local radio amateur, Thelma Morgan, K2OEW, broke in to invite us over to her place where her family had dozens of orange, grapefruit, and lemon trees. We went. We ate. Goodness, how we ate those delicious oranges; one after the other. The Morgans gave us a few boxes of them to take home too.

Well, the other part of this story has to do with my mother, who was in a nursing home at the time. Mom was unable to move or feed herself and was barely able to talk. When we went to visit her upon our return, we had a few of those fresh, hand-picked Florida oranges with us. Mom always liked oranges so my wife, Marie, peeled one and fed it to her in slivers. Honestly, the beautifully radiant smile that came over Mom when she tasted those delicious oranges still gives me shivers. She loved them. That memory of her smiling pleasure brought on by our now-famous "repeater oranges" will last forever.

I hope that I have convinced you of your future fun with repeaters because repeaters have become a major part of the amateur radio scene. They are very convenient and so effective. Show up at a *hamfest* (flea market for amateur radio equipment), and most of those around you will have an HT clipped on. Take a tour of the Empire State Building in New York City (as we once did), and you will find another ham up there working three states with a shirt-pocket walkie-talkie (as we did).

Almost every issue of every amateur radio publication carries at least one article on repeaters and related units. I wrote one on our local repeater station located high on top of the Playboy Casino in Atlantic City. We call it the "bunny broadcaster," "rabbit repeater," and a few other friendly names. I enjoyed writing that article. I enjoyed getting the background for it—they let me pose with one of their exotic bunnies. No problem. My wife, Marie, took the picture. The article, published in *CQ Magazine*, said, "Particularly appreciated was her [Marie's] photographic composition talents by which she several times directed the OM [me] to move in closer [to the Bunny]. Such are the trials of being a writer; if Cousteau can swim beneath polar ice caps for his material, can an amateur radio writer do less?"

I mentioned earlier that 2-m signals are usually line-of-sight coverage. However, when atmospheric conditions act up, we get *skip* (long-range radio-wave reflection) which brings in stations from hundreds of miles away. To prove the point, you remember reading about WAS—the Worked All States contest? Well, they've done it on 2 m also. In fact, some hams have powerful antenna systems on 2 m with which they wait for skip so they can contact a new state.

I am working on a higher-power 2-m station for the only purpose of being able to regularly contact my brother Leo, W3NWP, who is located in Frackville, Pennsylvania, a distance of about 150 mi. If he can reach halfway down to South Jersey and if I can get halfway up to him (and all we need is a bit more power than what we have), we will be able to have direct contact any time. The beauty of 2-m contacts is that only one station talks on the repeater at a time, so there is no interference. And it is a clear channel, free from atmospheric noises.

I remember sitting on the Penn State University campus in central Pennsylvania, talking to Leo from a distance of about 70 mi, and all this was with a little HT clipped to my belt. It was, as we say in the game, *armchair copy*, meaning it was as clear and as comfortable as talking over the telephone from an easy chair.

So much for the good news. Now the bad news. Novices are not permitted to operate on 2 m. You have to work for that privilege. It's just another carrot we hold out to entice you to upgrade after getting your novice license.

If you have enjoyed my personal review in this chapter of what repeaters can do for you, then take things one at a time—novice class first, general class to follow (license classes are explained in the next chapter). Many, many others have done the same thing; most of them knew nothing about radio before starting in for their novice licenses. They simply had the interest and desire. You should listen to how some of them excitedly talk about their fun in amateur radio—of which repeater use is one of their favorites.

# · 17 ·

# Licenses

In order to talk about amateur radio licenses, we need to first talk about the International Morse Code because the code is still required for all amateur licenses. So let me tell you a little bit about the code here, and a lot about it in a later chapter.

Perhaps more than anything else in amateur radio, our use of the code on the air sets us apart from the rest of the world. Code, you see, is a dying art these days. Everything is computerized and made automatic by the new electronic gadgets. Very few others use code on the air except in amateur radio. Depending on how you look at it, you might ask why we bother with code if it is a dying art. I won't go into my answer here except to counter with another question: Why bother to save the bald eagle from extinction? You see, if we radio amateurs gave up on code, it could soon become a lost art and forgotten in history. Besides, we really have as much fun in code operating as we do in voice operating.

Contrary to what you might first think about talking in code, it does offer as much of an expression of your personality on the air as your voice does. And it really isn't as slow a process as you might imagine.

You see, our code talk is sprinkled with common abbreviations such as agn for "again"; cul for "see you later"; tks for "thanks"; c (from the Spanish, "si") for "yes": tmw for "tomorrow"; and a whole bunch of others, all of which get to be second nature soon after you are operating on the air. (Some common abbreviations we use are listed in the appendix.)

Even more, we have dozens of internationally used Q signals which are recognized in all countries of the world. With these Q signals we can hold a short conversation with amateurs from any country in the world without knowing a thing about their languages. These Q signals (the more popular ones) are also listed in the appendix, but, to give you an idea of how useful they are, here are a few examples:

QTH?—What is your location?
QRL?—Are you busy?
QSY—Move to another frequency.

The abbreviations and Q signals make our conversations look like an ad in the classified section of a newspaper. But the codes enable us to pack a lot more into our conversations than we would if we had to spell out everything. We also develop a *telegraphic* style by which we don't need to send the obvious words (such as *is, and, but, of*) because they really are not necessary to communicate a thought or expression. All in all, I would estimate that an on-the-air code conversation at 20 words per minute is equivalent to a personal conversation with someone on the telephone at about 50 or 60 words per minute. That's not so slow, is it?

Another technique that helps us to move right along in sending code is something most of us develop after we've been using it for a while, that is, *empathy*, which is a fancy word for guessing what the other person is talking about even before they are halfway through a sentence. We get that way particularly with someone we've talked to several times on the air. Empathy is possible because code is not a cold process but is much the opposite, and loaded with personality. Believe me, code is unique as a means of conversation.

Learning code is not a stumbling block to amateur radio; not at all. One of the first steps to becoming good at it is to develop what we call the "code-happy" state of mind. Being code happy is when you bring your car to a stop and subconsciously spell out the sign in code: "Dit-dit-dit. Dah. Dah-dah-dah. Dit-dah-dah-dit." S-T-O-P.

Now, let's get into the different classes of amateur radio licenses we have.

First of all, and most important to you, is the novice class license, which is what this book is all about. Novice examinations require a code speed talent of 5 words per minute (WPM). As soon as you learn the code alphabet you are just about at a code speed of 5 WPM. For test purposes, a code word has five letters to it, so 5 WPM works out to only 25 characters (letters) per minute, or about 1 every 2 seconds. That's not too fast, is it? (Some code tests may require a demonstration on your part of being able to send the code, but that's no problem because it is natural that we are able to send code faster than we receive it.)

Written portions of the novice exam are thoroughly described in Chap. 18 on the novice license. Briefly, it will be 20 questions, either essay or multiple choice (it is up to the examiner as to which type will be used). Most of the questions are on rules, regulations, and operating procedures—what I consider to be nontechnical subjects.

Naturally, with the novice license being the first level of amateur radio licensing, the fact that the exam has a minimum of radio theory is matched by the fact that the novice class license also offers the minimum in operating privileges. After all, the novice license is intended to get you into the hobby so that you can get a taste of it and find out for yourself what ham radio has to offer, and to go on from there. That's why they call it a "novice license."

For example, the novice license does not allow use of voice transmission over the air, only code. That is good because, if all you can operate is code, then you are practicing code with every contact (best way in the world to practice code is over the air) and improving your code talents for the higher licenses. And all the while enjoying it.

I've talked a lot about DX (contacting foreign countries), and perhaps you may be wondering if novices can operate DX. You bet! Many novices have made DXCC (contacted 100 foreign countries). Some say they have surprisingly good luck at it because DX stations who want to get away from the crowded DX bands will tune in the novice bands, much to the delight of U.S. novices. Novices can get just about any of the award certificates around, plus they can take part (if they wish, of course) in traffic nets, Field Day, and honestly, just about all other radio activities permitted by the novice class license.

Novice operation is restricted to certain parts of the ham bands, as you learned in an earlier chapter. But novices are not restricted to talking only with other novices. Not at all. Experienced operators often drop into the novice bands to encourage the shaky hands sometimes found there. I do it regularly, and sometimes I get a surprised question, "What are you doing here?" Well, I enjoy it. I enjoy contacting beginners. Many of them are youngsters, and at least half of them are senior citizens, many of whom are retired and are learning what amateur radio can add to their lives in retirement.

All in all, the novice license serves as an apprenticeship toward the full range of "hamdom" which is at the next level, the general class level.

General class operation isn't quite the full bag of what amateur radio offers, but it is very nearly that. I would estimate that perhaps 95 percent of all radio amateur privileges come with the general class license.

You work harder to get the general class license—the code speed requirement is 13 words per minute, and the written test is more detailed in radio theory. But you get what you pay for, and paying attention to learning more about radio theory pays off many times over, believe me. Besides, being a novice and being exposed to these things makes learning general class requirements more fun than work.

General class license holders are permitted to operate on all amateur bands and in all modes—code, voice, amateur television, repeaters, radio teletypes (see glossary)—everything. Actually, the only difference between the general class licenses and other, higher licenses is that small parts of nearly every band are set aside for higher class licenses.

Although the code speed requirement for general class is 13 WPM, there is a special class of license called *technician class* which has the same code requirement of 5 WPM as the novice class but the same radio theory test as the general class. The technician class license operation is limited to code on the amateur bands below 10 m (30 MHz) but permits voice operation above 10 m (which includes use of the 2-m repeaters (see chapter on repeaters and on amateur bands).

Just because Mount Everest is there, people climb it. Just because an advanced class license is there, people go for it. If your skills and desires move you past the general class license, there are two more licenses—advanced class and extra class—each of which offers just a bit more of the pie. The *advanced class* has about 98 percent of all frequencies, and the *extra class* (the top radio amateur license) has 100 percent of all frequencies.

The advanced class exam does not require any higher code speed; it is strictly a technical exam on radio theory. The extra class, however, has not only advanced theory but also a required code speed of 20 WPM. About 8 percent of U.S. radio amateurs are extra class license holders; most (about one-third) are general class.

Radio amateur licenses are valid for 10 years, after which they must be renewed by application to the FCC. (There is a 2-year grace period, should you forget to renew. In other words, you have 12 years from the date on your license to renew it.)

A few years ago, the FCC started issuing call signs based on license class. As you read in Chap. 8 on call signs, there are certain letters, called *prefix letters*, assigned to the United States which come before the number of the district in which we are licensed. These letters are followed by the district number and by the *suffix letters*. Novices are generally assigned call signs with a two-letter prefix and a three-letter suffix (for example, AA#AAA); general class are single-letter prefix and three-letter suffix (for example, A#AAA); advanced class call signs are two-letter prefix and two-letter suffix (for example, AA#AA); extra class call signs are single-letter prefix and two-letter suffix (for example, K2VJ).

However, the FCC does not require that we change our call signs every time we upgrade our licenses to a higher class; it is entirely at our option. If we get a call sign we like as a novice, we can stay with it right up through extra class if we want. For this reason, it doesn't necessarily mean that someone we contact with a "two-by-three" call sign is a novice; they could very well be an extra class license holder who just didn't want to be bothered with changing his call signs. My son Jim, WA2JNN, is an extra class amateur who still uses his original novice call sign. After all, if you buy a bunch of QSL cards

(cards we swap with each other after a contact), those cards have your call sign printed on them, and to change means either handwriting your new call sign on them or having new QSL cards printed.

With all this talk of higher license classes, you might think that only engineers make it to the top; yet, you couldn't be more wrong. Of the more than 30,000 who are extra class hams, quite a few are husband-wife teams. Someone somewhere realized how many of these happy couples there were in the hobby and decided they should get to know each other better, so they formed a national club. With today's divorce rate of one divorce in every three marriages, you can believe that husband-wife teams in ham radio offer the togetherness that keeps everything close to home. These husband-wife teams are super couples in a super hobby.

Only a small percentage of radio amateurs are engineers. Most are in every occupation imaginable. I have the greatest respect and admiration for nonelectronics people who get into the hobby and eventually upgrade. Being in radio communications for so long, I found it easy to make my move up. I am absolutely impressed by the clergy, doctors, mechanics, secretaries, nurses, and so on, who took the time to better their lives through amateur radio. And kids too. My son, Jim, WA2JNN, passed the novice test at age 11 (in fifth grade); the general class exam at age 12; the advanced class at age 13; and I was mighty proud the day he walked out of the FCC office with the smile of victory on his face after passing the extra class exam at age 15.

Which brings me to another point about the amateur radio exams, one you will be pleased to know. In the past, we used to have to travel to the nearest FCC office in order to take the amateur exams. This often meant travelling for several hours or staying overnight in order to be ready for the FCC scheduled exam at 9:00 a.m.

No more, folks. Nowadays, radio amateurs all over the country can give the exams, not just the novice exam but all other classes of exams.

And they grade them right on the spot, so you know right away whether or not you passed. No waiting weeks like the rest of us used to.

# · 18 ·

# The Novice Exam

Good news! The Federal Communications Commission has changed the entire process of handling the novice class amateur radio test. Those changes will please anyone who wants to become a radio amateur.

One of the changes is that all questions (and answers) on the novice exam are now available for study. There is a pool of 200 questions from which volunteer examiners (the radio amateur who gives you the test) selects 20. Those 200 questions are public information.

You no longer need wait to find out if you passed the novice test. Your examiner will grade your test on the spot. Your passing grade (15 or more correct answers) is sent to the FCC, who issues you a license and a call sign.

You can take the novice test in the comfort of your own home or anywhere your examiner agrees to. I have given the novice test several times at our kitchen table, which seems to be a relaxing place. At least everyone to whom I've ever given the novice test has always passed. Further, you can make arrangements to take the test at a convenient time such as at night or on weekends.

Any radio amateur (age 18 or over) licensed at the general

class level or higher can give the novice exam if he or she is not a relative of the person taking the exam.

In Chap. 23, "How to Join," you will read about ways to make contact with a local radio amateur or volunteer instructor. You will need someone to give you the exam, even if you do not need help with your studies.

I will talk quite a bit about the novice written test in this chapter. And since the novice test is in two parts, code and written, I will also talk about the code requirements. Then I will separate the novice written test into the nontechnical parts and the technical parts. I think that when you are done with this chapter, you will be surprised to find that things are not what they may seem to be. (Remember the lion?)

## CODE

I explained earlier about learning code in the chapter on licenses, and how it helps if you become code happy while you are learning. In a later chapter I will list several of the cassette code practice tapes you can buy to help you learn the code.

For now, however, I would like to do some motivating to convince you of why you should learn the code.

Is learning the International Morse Code so difficult?

Not for me, it wasn't. I advanced to a speed of 18 words per minute (WPM) in 2 months, but I practiced every day. I have since taught the code to many people, and I've got a secret about learning the code that never fails. I'll pass that secret along, very soon.

The novice code speed requirement is 5 WPM. At an average of 5 letters to the word, 5 WPM is less than 1 letter in 2 seconds.

Practice is what does it. Daily practice, even 10 or 15 minutes at a time. By practicing daily, you condition your mind to appreciate that you are learning a new language. That's not my secret, but daily practice is a big part of learning code.

The experts have come up with new ways to learn the code, ways that I find to be much better than the old ways by which I first learned. They say you should never write out the dits and dahs in a code letter because that brings your sense of sight into the act, which is a distraction to learning. (That's not my secret, either.)

What you want to do, instead, is to learn code strictly by its sound. Never look at the code letters of the alphabet written out in dits and dahs. Stay with learning by sound. You want to learn code through a process of immediate association between sound and hand, ear and mind. Bringing in the sight sense can slow you down when you try to build up to a faster code speed. Learning by sight might seem easier at the beginning, but it is the worst way as you move up to higher speeds.

Another modern improvement they have come up with for rapid learning of code is to send code practice to you at a letter speed of 13 WPM instead of the 5 WPM required for the novice test.

The advantage to this technique is that the mind tends to hear a code letter differently when it is sent slow as compared to when it is sent fast. If the letter is sent the same speed (13 WPM) when you first learn, when you advance in speed it will sound just the same except with shorter time gap between letters. The result is that you have no hangups in increasing your code speed if you learn the right way from the very beginning. As far as I know, all commercial code tapes use this fast-code method. (Nor is this my secret for learning code.)

"Why in the world do we need to learn the code?"

That's a question I get all the time. We all have our own answers too. No-code licensing is a sensitive issue these days. The FCC wanted to issue a no-code license that had a stiff technical requirement in theory and computers but granted limited operating privileges. I can only give my own answer to the question, which may not be the same as anyone else's.

First of all, code is required for the novice test, and that's a fact. It's going to stay that way for a long, long time too, so there is no use wishing it would go away. Part of my secret to learning code easily is to accept the fact that you must learn the International Morse Code in order to become an amateur radio operator.

We radio amateurs are probably the very last in a long line of International Morse Code operators in the world. Even seagoing ships are doing away with their *sparks*—shipboard radio operators—in favor of *communications officers* who now handle all the modern electronics systems onboard seagoing cargo ships. Being last in a long line of something special gives us an inheritance to hold on to.

One of the arguments we make is that code will get through under difficult radio conditions when voice will not, but that isn't entirely true. What is true is that code has a better chance of getting through under the same conditions as voice, and that is always true. After all, you need only copy a single tone for code but you must get all the sounds for voice in order to understand what is being said.

Another point is that we can fit three code conversations into the same radio frequency space as one voice conversation. That's important. Think of what the bands would be like if all of us used voice at the same time. We'd be in trouble with interference all over the bands.

Now, to my secret about learning the code. This has to do with a story I like to tell about when I was a shipboard radio operator with a message to send back to New York while we were in the Persian Gulf. When I fired up the radio, I found the ship-shore code bands to be just about dead. (We had only code transmitters on board ship in those days.)

I called the New York shore station, and after a while I heard a very faint reply. It was enough for me. I won't go into detail, but I can tell you I spent the next 3 hours trying to get my 10-word message through, and we used as much mental telepathy as we did radio telegraphy. I think I had lost 5 pounds in the 110-degree weather we were in, and my ears ached for days afterward from pushing the earphones tighter into my ear, trying to pick out the very faint signal from the New York station. I still remember that day quite well. I still like to brag about it as some real radio operating.

Well, if you think I told that story at this point just to show that code got through when voice would not, you are mistaken. No, the moral of my story has to do with my personal justification for keeping the International Morse Code alive, and that is in the satisfaction, pleasure, challenges, and memories which code operation seems to so often produce, particularly on the ham bands.

I liked the code that day I sent my 10-word message in 3 hours; I like it even more today. Which is the bottom line to where I hang my hat in the discussion of no-code licensing, and it is also my secret for learning code: Do not simply learn it, *enjoy it!* Enjoy it, and you will learn. That's a promise.

## NONTECHNICAL

The new novice written exam is in nine areas of study: rules and regulations (7 questions given on the novice exam); operating procedures (1 question); radio-wave propagation (1 question); amateur radio practices (3 questions); electrical principles (3 questions); circuit components (1 question); practical circuits (1 question); signals and emissions (1 question); and antennas and feed lines (2 questions). These add up to 20 questions, of which you must get 15 right in order to pass.

Some of these subjects may seem complicated, but they really aren't. Of these subjects, I would say that the rules, regulations, operating procedures, and amateur radio practices are nontechnical; that is, you don't need to know about electricity or radio to handle them, yet they make up the bigger half (11 questions) of the exam.

Novice class rules and regulations are simply common sense, as you will see by a few examples. All the novice study books have the subject covered quite well. These books are available from the *ARRL, 73 Magazine, CQ Magazine*, Ham Radio Bookstore and most of the amateur radio supply stores.

You can best learn about rules and regulations by first learning what ham radio is all about, which is exactly what you are doing with this book. If you start off with a good background in the hobby, you will then understand rules and regulations much more easily. For the written exams, your volunteer examiner has the choice of giving multiple-choice questions: fill-in-the-blank (single-answer) questions; or write-out-your-answer (essay) questions.

I personally think the only good thing about essay questions is that if you write everything you can think of on the subject in question, sooner or later you might stumble across the right word and can argue that your answer is right. On the other hand, essay questions are tough to grade. I know; I've handled quite a few in my time as a correspondence course instructor.

With multiple-choice questions you can always take a guess at the right answer. And they are easy to grade, with no chance of argument. Usually. The only problem with multiple-choice questions has been with the examiner. Unless an examiner keeps a clear record of all questions that are missed and is able to determine that a given question (or answer) is bad if everyone misses it, then the multiple-choice exam will not work as well as it could and should.

Part of the problem with "bad" multiple-choice questions is not with the questions, technically, but in the use of unnecessarily complicated English—big words, I mean. We still have that problem with the FCC amateur radio questions. (I'd be pleased to offer my assistance to rewrite their questions in plain English.) You will see what I mean about plain language in some of the FCC questions I will be quoting later in this chapter.

Here are a few of the questions from the first set to the novice exam. I like these first few because they are rather plainly written.

1. What is the amateur radio service?

*Answer:* The definition of the amateur radio service needs explanation at this point. You see, amateur radio is defined by law as: "a means of radio communications between amateurs; a radio communications service of technical self-training; and a means of technical development." That's just about how we are carried in the books, and with good reason. With the commercial services all over the world always sniping at us, trying to get more and more of our frequencies away from us, we need some formal definitions of our purpose in order to carry weight when our radio ambassadors sit down to negotiate. We did all right the last time. We walked away with three new bands. We never would have if our only purpose for existence, by law, had been to have fun talking over the air. Public service and emergency preparedness are a big part of our continuing existence.

2. Who is an amateur radio operator?

*Answer:* Anyone who is licensed to operate in the amateur service.

3. What is an amateur radio station?

*Answer:* A radio station that is licensed by the FCC to operate in the amateur radio service as defined by law.

4. What is amateur radio communications?

*Answer:* An amateur radio communication is a noncommercial amateur radio conversation over the amateur radio bands (we call it a QSO—see the glossary).

5. What are the novice class operator transmitting frequency privileges in the 80-m band?

*Answer:* (I really wish they had used simpler language. All they are asking is, do you know what frequencies novices are allowed to operate in the 80-m band?) The answer to this question is given in Chap. 7, "Amateur Bands." Thumb back for the answer because I gave that information both in the text and in a nice illustration, just to address this question. By the way, they also ask this same question for the 15-m and 10-m bands, so one of them is very likely to be on your exam.

6. What is the only emission authorized for use by novice class operators?

*Answer:* (Translation: Are novices permitted to send by code or by voice?) You know the answer to that—code, of course.

7. Under what circumstances, if any, may an amateur radio station be used to transmit messages for hire?

*Answer:* Never. Friend, if you ever dare to take money for sending an amateur radiogram, we'll jump on you. We are strictly nonprofit and noncommercial in what we do over the air. Don't ever forget that.

8. In what call sign district will your amateur radio station be located?

*Answer:* This depends on where you tell the FCC your station will be located. You can answer this question by checking App. 4 on call sign districts.

9. How do I identify my amateur radio station communications?

*Answer:* You (station licensee) send your call sign at the

beginning and end of a conversation and at least once every 10 minutes during the conversation.

10. What does S in the RST signal report mean?

*Answer:* You can look up this answer in App. 5, "Sample Code Conversation, QSO," which also has a list of the more popular code signals we use on the air.

11. To protect against electric shock hazards, the chassis of all equipment in an amateur radio station should be connected to what?

*Answer:* The chassis should be wired to a good electric ground such as a rod driven into the ground.

There is your sampling of the nontechnical questions. Perhaps you are beginning to see that learning this part of the test is little more than common sense and that it will all be made much easier by reading this book.

## TECHNICAL

Writing this part of the chapter is a personal pleasure that comes from my past as a technical writer and correspondence course writer. From that background, I've learned a few things about how to present technical material in ways that are most likely to be understood by readers, and not necessarily to impress them.

The material to follow is a quick course in electricity; it is certainly not the whole course. I want only to introduce a subject without suggesting you have to learn everything I talk about. What you learn here could be valuable to you in your later novice studies, whether that be on your own or in a formal novice course. (Someday I plan to write the entire novice course, and it will be quite a bit like what follows.)

When you study electricity, one of the first subjects you always get into is Ohm's law. Even though there is not a single question on Ohm's law in the novice test, my judgment is that you must know something about it in order to learn the basics of electricity and radio. Therefore, I'm going to use a few pages to talk about Ohm's law.

My dictionary defines Ohm's law as: "In an electrical circuit, the current in amperes is directly proportional to the electromotive force in volts, and inversely proportional to the resistance in ohms."

Whew, that's a mouthful. Fortunately, I have better ways of saying the same thing. Just stay with me.

What my dictionary says is true, but it could have said it in less stuffy terms. After I have explained Ohm's law my way, I'll give a better definition, one that can be more easily remembered.

My explanation of Ohm's law has to do with an outdoor water pump, the kind that sits outside the old farmhouse. When you pump away on the handle of a water pump, the force from the handle builds up a pressure that makes water come up through a pipe that runs down into the ground. The pump is a force that makes water flow through the pipe.

A force is also needed to make electricity flow through wires in an electric circuit. What kind of force, and where does it come from? A battery is one example of an electrical force. The outlets in your house also have an electrical force that comes to you from the electric company's generators. The battery and those generators are both electrical forces. They force electricity to flow through the wires that connect them. This is much the same action as we have at the pump when pressure is built up and forces water to flow through the pipes. In electricity, the electrical force makes electricity flow through wires.

We call the battery and generator *electromotive forces* because they force electrical motion in a circuit. And now, suddenly, one of the technical terms to my dictionary's complicated definition of Ohm's law is explained: "Electromotive force is an *electrical* force that causes *motion* of electricity in a circuit."

Electromotive force is measured in units of *volts* (named after Italian physicist Allesandro Volta). Batteries are rated in volts. For example, car batteries are 12 volts; transistor radios are usually 9 volts; flashlight cells are 1.5 volts; house power is 115 volts. (The abbreviation for volts is V.)

When water flows through the pipe, we call it a *current flow* because the water flows in a current. When you pump on the handle several times, you get a flow of water up through the pipe and out the nozzle. You got that current flow because you applied a force to the pump handle. It's the same way in electricity.

When you apply a force—an electromotive force—to an electric circuit, electricity will flow through the wires of the circuit. Water flow is measured in gallons per minute; electricity flow is measured in *amperes* (named after French physicist André Marie Ampère). Ampere is abbreviated as A and is commonly called *amp*.

The flow of electricity through the wires of a circuit is very much like the flow of water through the pipes of a pumping system. In both cases they are called *current flow*.

Let's stop a minute to go over these new terms:

1. *Electromotive force:* A battery or generator which provides voltage in an electric circuit
2. *Volts:* The unit of voltage produced by electromotive forces
3. *Current flow:* The flow of electricity in an electrical circuit to which an electromotive force has been applied
4. *Ampere:* The unit of current flow in an electrical circuit

This is not a bad beginning toward understanding Ohm's law. Stay with me.

The third term in Ohm's law is *resistance.* I will spend more time on this term because resistance is not as familiar to most people as are volts and amperes.

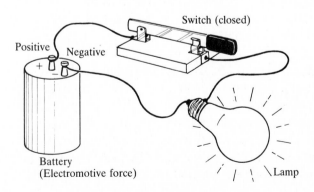

Normal electrical circuit.

Again, back to the water pump. This time, imagine we have two identical water pumps, except one is fed by a 1-in-wide pipe and the other by a 4-in-wide pipe. For the same number of pumpings on the pump handle, which pump would you expect would deliver the greater current flow? The 4-in pipe. Why? Because the 1-in pipe, being narrower, has much more resistance to the flow of water than does the 4-in pipe. It's the same reason why a $^5/_8$-in garden hose puts out more of a current flow than does a $^1/_2$-in garden hose. The resistance of the smaller pipe-hose is greater than the resistance of the larger pipe-hose. It's the same in electricity.

In two otherwise identical electric circuits where one has a resistance greater than the other, the circuit with the higher resistance will have less current flow than will the circuit with the lower resistance. In other words, more current will flow in a circuit with low resistance than in a similar circuit with high resistance. That statement is half of Ohm's law, but maybe you don't yet recognize it that way. You will soon.

It is important for you to know that we have resistances in everything that has an electric circuit—toasters, hair dryers, radios, electric heaters, air conditioners, electric ovens, fans. In fact, we make electrical parts that are called *resistors* which we make very good use of in electric circuits to control the level of current flow.

Why should we want to deliberately resist the flow of current? Simple. If we didn't have resistance in a circuit, the fuse would pop. We would have a *short circuit.*

An example of a short circuit is when a wire is connected directly across a battery. A jumper cable right across your car battery would be a short circuit. A short circuit has zero resistance in it. And with zero resistance in a circuit, can you guess how much current will flow? A huge amount. Remember, the more the resistance, the less the current; the less the resistance, the more the current. And if you have zero resistance, good-bye battery; it will burn up and the jumper wire will get red hot. Take my word for it; don't try this on your car battery. We put fuses in circuits to protect the circuit against a short circuit. (When the fuse pops, we get an *open circuit.* I will explain open circuits shortly.)

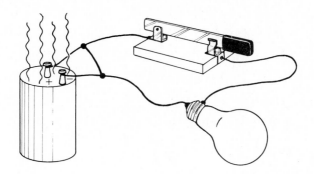

Short circuit.

A short piece of wire doesn't have exactly zero resistance, but it has a very low resistance. (I did say that everything has resistance to it.) Suppose, though, that we had an electric circuit 2 mi long. If we connected such a long run with thin wire, it would have a lot more resistance than if it were run with thick wire. And the thin wire would have less current flow than would the thick wire. It's the same as the two pumps with the thin and thick pipes.

Another example of what resistance is can be found in an ordinary light bulb. Bulbs are made with a special metal inside such that when the right voltage is connected to the bulb, it will heat up to where it is white hot. That's how we get it to make light.

The lower the resistance of a bulb, the more current will flow through it. Then, which has a lower resistance: a 60-W bulb or a 100-W bulb? The answer is the stronger-current, lower-resistance, 100-W bulb.

Let me now restate Ohm's law in my own words: "In an electric circuit, the current flow increases when the voltage increases, and it decreases when the resistance increases."

We've spent some time convincing you that the current decreases when the resistance increases. I think you know that if you apply a greater voltage (force), that the current flow (amperes) will also increase.

When you think about it, we could almost write Ohm's

law for a water pump by saying: "In a pumping circuit, the current flow increases when the pumping force increases, and it decreases when the pipe size decreases."

I think we're ready for Ohm's Law in formula form:

$$R = \frac{V}{I}$$

where $R$ = the symbol for resistance in ohms (named after German physicist Georg Simon Ohm and abbreviated $\Omega$)
$V$ = the voltage (electromotive force) in volts
$I$ = the current flow in amperes

The way you read this formula is: "The resistance in an electric circuit is equal to the voltage across the resistance divided by the current flowing through it."

The following table summarizes Ohm's law:

| UNIT OF MEASURE | SYMBOL | UNIT OF ABBREVIATION |
|---|---|---|
| Volts | $V$ | V |
| Ampere | $I$ | A |
| Ohms | $R$ | $\Omega$ |

A few example problems make all this clear.

Suppose we had a lamp that plugged into the house outlet, which is 115 V, and say the bulb in the lamp draws 1 A of current when it is lit. What is the resistance of the bulb? Use Ohm's law to figure the answer:

$$R = \frac{V}{I} = \frac{115}{1} = 115 \ \Omega$$

As a point of interest at this time, we learn to say things such as: The bulb has 115 V "across" it; 1 A "through" it; and 115 $\Omega$ "in" it. Small point, but saying it right, this way, lets others know you know something of what you are talking about.

Try another problem on Ohm's law: The dome light of a car draws 1 A when lit. Car batteries are 12 V. What is the resistance of the dome light?

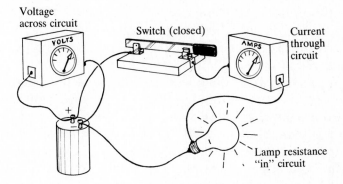

Voltage "across," current "through," and resistance "in" a circuit.

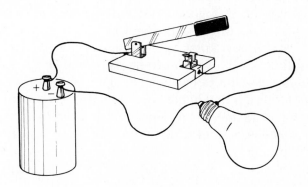

Open circuit.

$$R = \frac{V}{I} = \frac{12}{1} = 12\ \Omega$$

Let's get back to the open-circuit condition I mentioned earlier. An open circuit is what you have when the switches in the house or the headlights in your car are turned off. Your switch opened the circuit; therefore, we call that an *open circuit,* whether it is by turning a switch off or by a wire that is not connected to complete the circuit. The important thing about an open circuit is that no electricity can flow through it. After all, an open circuit is an infinitely large resistance and, as you can figure from Ohm's law, an infinitely large resistance will cut off the current flow entirely.

An open circuit is the opposite of a short circuit. In the open circuit, no electricity will flow. In a short circuit, a large current will flow.

We're just about done with Ohm's law at this point, but we're now about to learn a few things that involve Ohm's law.

For instance, wires are made of metal because metals have very low resistance. We want very low resistance in the wires that connect between parts in a circuit because, for example, if a line cord has a high resistance, it will rob power from the lamp. A lamp cord should be low resistance. Metals are good conductors of electricity because they offer low resistance to the flow of current. Examples of good conductors are copper, aluminum, gold, silver, and many other metals.

The opposite of a good conductor is what we call an insulator. Insulators offer a very high resistance to the flow of current. As you now know from Ohm's law, when we have a high resistance, very little current will flow. If we have an infinitely high resistance (such as the open circuit), no current will flow. Materials that make good insulators are wood, glass, rubber, porcelain and air, to name a few. These materials are all poor conductors of electricity; they are good insulators.

But nothing is perfect. Although insulators do not normally allow current to flow, if the voltage across them is in-

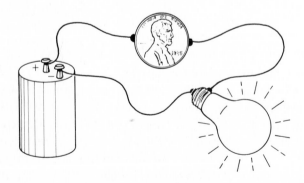

Copper is a good conductor.

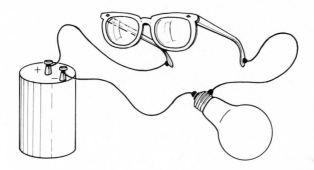

Plastic and glass are insulators.

creased to a very high level, the insulator will eventually break down and will then allow a current to flow through it. Lightning is a good example. The insulation (air) breaks down and lets huge amounts of current flow through the air when lightning strikes. The voltage of a lightning strike is in millions of volts. This is its "breakdown voltage." All insulators have a breakdown voltage. Breakdown voltage is defined as the lowest voltage that will cause an insulator to break down and pass a current through it. You will soon see why I mention this. Stay with me.

You will also soon see why I bother to mention that there are two kinds of electricity—dc and ac. The abbreviation dc means "direct current." You get dc from a battery. Electricity from a battery flows out of one battery terminal (which is often called a *battery pole*), through the circuit, and over to the other terminal. It goes *directly* out in one direction. It is called "direct current," or "dc."

House current is different. It goes out and back, out and back, 60 times a second. It alternates (cycles) back and forth, over and over again. It is called "alternating current," or "ac."

*Remember:* Batteries are dc; house current is ac.

Something else to remember is that a battery has two *polarities*; a positive and a negative; plus (+) and minus (−). You have seen those markings on transistor batteries, I'm sure. Now you know what they mean, if you didn't already know.

Direct current flows in one direction only.

Alternating current flows back and forth.

Now, then, we've sure packed a lot of information into this chapter. Here is your chance to see how much of it you understood. Here is also a chance to see the actual exam questions in the electrical principles part of the novice exam. What you miss here means you need more practice, so perhaps you should go back over this chapter some other time.

1. Electrons will flow in a copper wire when its two ends are connected to the poles of what kind of source?

    (a) electromotive or voltage     (c) reactive

    (b) donor     (d) resistive

2. The pressure in a water pipe is comparable to what force in an electric circuit?

(a) current
(b) resistive
(c) gravitational
(d) voltage

3. What are the two polarities of a voltage?

(a) right-hand and left-hand
(b) forward and reverse
(c) positive and negative
(d) clockwise and counter-clockwise

4. What type of current changes direction over and over again in a cyclical manner?

(a) direct current
(b) alternating current
(c) negative current
(d) positive current

5. What type of electric current does not periodically reverse direction?

(a) alternating current
(b) periodic current
(c) direct current
(d) positive current

6. Which answer includes four good electrical insulating materials?

(a) glass, air, plastic, porcelain
(b) glass, wood, copper, porcelain
(c) paper, glass, air, aluminum
(d) plastic, rubber, wood, carbon

7. Which answer lists three good electrical conductors?

(a) copper, gold, mica
(b) gold, silver, wood
(c) gold, silver, aluminum
(d) copper, aluminum, paper

8. What is the term for the lowest voltage which will cause a current in an insulator?

(a) avalanche voltage
(b) plate voltage
(c) breakdown voltage
(d) zener voltge

9. What term describes the electric circuit failure that causes excessively high current?

(a) open circuit
(b) dead circuit
(c) closed circuit
(d) short circuit

10. What term describes an electric circuit in which there can be no current flow?

(a) a closed circuit
(b) a short circuit
(c) an open circuit
(d) a hyper circuit

**Answers**

1 (a); 2 (d); 3 (c); 4 (b); 5 (c); 6 (a); 7 (c); 8 (c); 9 (c); 10 (c).

# · 19 ·

# FCC Gettysburg

A while back, prior to Public Law 97–259 which changed a lot of the ways in which the amateur exams are handled, I had talked to some officials at the FCC in Gettysburg, Pennsylvania, to get an idea of what happened to the completed novice exam when it was sent back to them.

Well, so much changed in 1983 that what follows in this chapter is mostly history. However, the chapter does have some interesting points which might help convince you there are others out there with the same ambitions you have.

The FCC gets about 30,000 novice exams each year, of which 90 percent are successful and result in licenses being issued. Of the remaining 10 percent, one-fourth of them were rejected simply because the exams were not returned within the time period the FCC had allowed between when they sent the exam and when it should have been returned to them. That time period was 30 days, but it was increased in 1983 to 60 days. (Now it doesn't matter because the FCC does not issue exams anymore.)

Although the volunteer examiner now grades the novice exam, the Gettysburg office used to do this chore. A passing grade of 75 percent was required. The Gettysburg office then sent the information to Washington where it was (and still is) entered on a computer which generates a call sign according to the licensee's location. The license is then mailed to the applicant.

The time period for this process would vary from 2 to 6 weeks, depending on the workload at Gettysburg. Usually, within 4 weeks the license would be in the mail. It used to be said that if you didn't hear from the FCC in 2 weeks that you obviously passed the novice test. Now you know when you take the test whether you passed; all you wait on the FCC for, these days, is your own special call sign so you can get on the air.

Novice exams received by the FCC seem to peak on a daily basis, with Monday being the heaviest load. Although the daily average runs to about 150 exams, Mondays have twice as many as other days. Sometimes 50 or 60 exams are received from a single examiner who may have just finished a novice course.

Exams seem to peak by the season too, with the largest numbers of them being given between January and May.

The FCC keeps no official record of unusual ages of novice applicants, but you will read about the youngest, Guy Mitchell, WD0DVX, who was licensed at age 5. At the other end of the scale, Bill Welsh, W6DDB, who writes a novice column for *CQ Magazine* and has probably graduated more novice radio amateurs (thousands) than any other instructor in the world, once told me that a 92-year-old who took his course became a novice amateur radio operator.

All FCC field offices are listed in App. 2. If you ever feel the need for a more immediate contact with someone there, here is a special address and phone number for the FCC at Gettysburg:

> Federal Communications Commission
> Consumer Assistance Branch
> Gettysburg, PA 17325
> Phone: (717) 337–1212

# · 20 ·

# Equipment Costs

This chapter on typical costs of ham radio equipment might also have been titled, "Only the License Is Free." But the real question of interest to everyone is exactly how much money we're talking about. As you will see, it really isn't much for what you get in return.

The cost of brand-new commercial ham radio sets can run from a few hundred dollars up to a few thousand dollars. But, as a novice, who needs new? Take my word for it, ham radio equipment costs should not be a reason for holding you back. Besides, cost alone is not the measure of the hobby. With my past descriptions of hobby benefits, you can read through this chapter to help you make your own cost-benefit decision about the hobby.

How does $25 for a complete novice station sound? It's been done, and written about in the ham magazines. But that is an absolute bottom limit situation which I'll later explain.

Ham gear prices fit into three levels: (1) *low end*, meaning older, used equipment; (2) *middle line*, meaning moderately priced new equipment; and (3) for those who can afford it, there is the *upper line*, as there is in everything, isn't there?

The upper line is the supersophisticated equipment with all the bells and whistles.

There are two types of older used gear. One type consists of the old radios from back in the 1950 to 1960 period which don't have the modern features. They are mainly useful for code operation, which is all the novice cares about anyway. The second group is the reasonably modern (about 10 years old) used equipment. Let me explain a bit more.

"Back in my time" is an expression my son has come to dread hearing me say when I get into my spells of talking about the "old days." Back then, ham radio sets were separate units— a receiver on which to receive signals and a transmitter with which to transmit.

Modern rigs, however, are what they call *transceivers* in which the transmitter and receiver sections are combined in a single cabinet (such as in CB sets).

The old "separates" are super for code use. They were built of iron and weigh a ton. "Boat anchors" is what we call them. I picked up a boat-anchor transmitter (weighed about 60 pounds) and made some modifications to it which I wrote into an article I sold to *73 Magazine.* A case of fun and profit.

Plenty of those old sets are still around. About the best place to buy them is at one of our numerous *hamfests,* which every ham community has at least once during the year. They are otherwise known as *flea markets.*

"These old rigs certainly are heavy."

Some hamfests are quite large—40,000 people attend from almost every state in the country. Most are regional affairs with one or two thousand hams attending. We put up our used gear for sale by *tailgating*—that is, we just open the tailgate of the stationwagon or set out a card table and we're in business.

I did a survey of a few hamfests in the New Jersey–Pennsylvania area last year and came up with prices as low as $15 for a code transmitter and $30 for a code receiver. Better sets sold for up to about $75. I used to think it took about $150 for a complete novice station, but I have now revised that price down to under $100. Add $5 for a key and $5 for the antenna wire and you're in business.

Of course, you need a ham friend to do the shopping for you. You will read in a later chapter that one of the best moves you can make, to get into ham radio, is to have a friend in the hobby. I'll tell you how to go about finding a ham friend in that chapter.

Now, about the case of the individual who put together a complete station for a novice friend for $25. He did exactly what I just mentioned; he helped his novice friend shop at hamfests for some ancient boat anchors.

He did a little servicing on them because at those prices they needed a little attention. But he did it, and he proved that it could be done. I think the $25 may be an exception, and that $100 is the rule. So much for boat anchors (which are the most fun and greatest memory builders of them all).

In the same list of used equipment are the more modern transceivers which are going for bargain prices these days because what happened to CB has also happened in ham radio except on a much smaller scale—they expanded the bands on which we can operate.

A couple of years ago the international group on radio communications authorized three new ham bands. Well, you can see what happened—the gear then (and still) in use got to be slightly obsolete because it didn't have the three new bands. (Remember what happened when they upped the CB channels from 23 to 40? All the old 23-channel sets went down the drain and sold for peanuts. A lot of manufacturers went bankrupt from that move. It isn't quite that bad in ham radio, however.)

Thousands of those "obsolete" sets are still in use. I bought one for $250, a set which originally sold for $595 only five years earlier. It is a beautiful set with most of the modern features, except it doesn't tune the three new amateur bands. So what? It has everything else I need. It has all-transistor circuits except for the power output stage, which uses vacuum tubes. That's no problem either, as those tubes last 5 to 10 years depending on how much they're used. The set itself is so beautifully built that I expect it will last another 15 or 20 years. And get this— it will do absolutely everything the new sets do except tune the new bands and a few other toots and whistles such as built-in scanners and little computers to help you switch to a new frequency in a hurry—lazy man's stuff, I call it.

So there you have it—for $100 you can put up a complete novice station of boat anchors. Or, if you shop carefully, you can buy a reasonably modern and sophisticated used transceiver for $300; less, if you get as lucky as I did; and getting less each year as more and more new models come out to bring down the price of the older models.

I might add that the more sophisticated used equipment can be bought at half their original selling price of $800–$900. I use one, myself. I paid the old price (the high one, of course) and would now have to sell it for about half what I paid.

All in all, opening those new bands did you newcomers a big favor. But that's not all in your favor: As the manufacturers keep putting in more and more fancy improvements such as more transistors, including the power stages, scanners, memory circuits, and all that, it's enough to make the more experienced hams drool to where they just have to sell their perfectly good sets to get one of these fancy new ones. Their trade-ins are your gain. Ham gear never offered so much for so little as it does today.

Now we have those new sets to talk about. They are the full-line transceivers which come in low, middle, and upper lines. You can buy a few models under $500, but most of them are in the $700–$900 class. They have all you need for a dozen years to come. I suggest you look them over carefully and not be swayed by the fanciness of the higher-priced jobs.

In the middle line, prices range from $1000 to $1500. I've

been looking over a few of those figuring, well, my years of operation and my activities say I'm entitled to a little better than what I have. So far I've resisted, and I expect I just might hold off for some time. Besides, all those computer-aided circuits mean I've got more to learn just to turn the set on. But they sure are tempting. Most of them have digital displays (numbers) to show the exact frequency you are on.

And at the highest end of the line? Shucks, if that's where you live, just get any ham to help you pick out one. Most of us would drool over getting our hands on one of them, so there is no problem getting advice on how to lay out a few thousand dollars.

A feature useful to some hams with regard to these newer sets is the fact that most of them work in the car just as well as in the house. That's because they are transistorized and are powered from the 12-V car battery. You can use either the built-in power supply (power converter to convert from house voltage of 115 V ac to 12 V dc—see glossary), or you can use a special power supply for home use. Most sets now come with a cable which connects to your car battery. You then mount the radio in the car, hook a mobile antenna on the back bumper or on top, and you're in business.

Of course, that's not yet of too great concern to you, as a novice, since you're restricted to code use. (I've heard a few diehards on the air who operate code from the car. Tricky business, that. Actually, I'd like to try it myself someday.)

Mobile, by voice, is a big feature to amateur radio. Newer sets are no bigger than CB sets, yet they tune all the ham bands and put out over 100 W. An all-band mobile rig makes a fabulous companion to those who are on the road a lot. Imagine regularly talking to Europe from the car while on the way to work in the morning. It's fun. And most of us extend an informal priority to mobile stations which is not a requirement by law but is, instead, a matter of courtesy. Most nets conduct their roll calls by first calling for any mobile stations wishing to check in.

So far, I've been talking only about HF sets (those that cover from 160 m to 10 m). After building you up in a previous chapter about how fabulous 2-m repeater operation is, let me now talk about the cost of 2-m walkie-talkies and home sets.

Walkie-talkies as used in commercial operations such as police, fire, security, and the like usually cost $1000–$2000. Walkie-talkies for 2-m amateur use cost only a little over $200. The commercial sets don't do anything more than do the ham radio sets; in fact, they do considerably less. We have touch-tone pads (for telephone hookup), scanners, wide frequency tuning range, and more. For about $300 you can get a ham radio walkie-talkie that has a computer chip that provides you with memory of your favorite repeaters, provides band scanning, figures out your paycheck, salutes you in the morning . . . I'm kidding. But I'm not kidding when I say that for $300–$350 you can get an HT that outshines any of the commercial kilobuck ($1000) sets. Believe me, you get much more than $200 worth of fun and pleasure from a ham radio walkie-talkie.

Two-meter base (home) stations range in price from $75, used (not too many features), to about $350, new. They are built so that they can also be used in the car. There probably are more 2-m mobile operators in the country than all other mobile operations put together, including the commercial services. That's how popular 2-m mobile is.

Worried about having your ham gear stolen? First, you should declare it separately in your house insurance. One fine way to prove positive ownership is to take a few pictures of it (and you) in your ham station. We all do. That photo will support any theft claim.

Second, the ARRL sponsors an equipment insurance program, which they will tell you about on request.

Also, the National Crime Information Center's computer in Washington, D.C., supposedly has our call signs stored in its memory banks. I've been told that should your stolen equipment be recovered, they'll find you quicker if you inscribe your call sign and your two-letter state abbreviation on the equipment. For example, with my call sign of K2VJ and living in New Jersey, I put NJK2VJ on my equipment. This is called the "Owner Applied Number System." I don't know how it works, but it is worth the effort.

We've been talking about radio sets so far. You also need

an antenna to get on the air. Ham band antennas are of two types: *wire antennas* (inexpensive) and *beam antennas* (expensive). Let's first talk about beams.

My beam antenna is a huge job that stretches out 24 ft long by 31 ft wide. Measure that length and depth in your backyard and then picture it up in the air. It's big. Some hams have even bigger beams.

I bought that beam 7 years ago at a bargain price of $200. It sells for almost twice that now. I bought it as my son Jim's reward for having passed his advanced class exam.

But the beam wasn't the whole of it. No way. You need a tower to get that rascal up in the air where it will do some good. Tower prices are out of sight. New, they run from $5 a foot up to $25 a foot, depending on how fancy you want— straight-welded or bolted job, or a tilt-over or telescoping tower so you won't have to climb it.

My compromise was an old 40-ft steel tower, which we scraped and repainted. It looks good. If and when you are ready for a tower, search around for someone with an old TV tower who might like to have it taken down just to get it out of sight. That happens.

You need something by which to swing a beam antenna around. That's the name of the game with a beam—to beam it at the station you are talking to for best results. Rotors for ham beams cost much more than do rotors for TV antennas because they need to turn much heavier loads. Ham rotors cost from $125 to $400, new; maybe less for used ones, but you need to buy a used one where you can first test it.

All in all, the price of a complete beam setup can run as much as the price of a lower-end new transceiver. Some say it is worth it. You can decide for yourself when you advance to higher class licenses. Our beam is the first I have ever owned in all my years of ham radio.

What about transmitter power? The legal amateur radio transmitter power limit in the United States is 1000 W for all classes except novice, for which the legal limit is 250 W. By comparison, CB sets run 5 W. Most modern transceivers are in the 150-W class.

A surprising number of hams are dedicated to using low

power of less than 5 W and less than 1 W. Their kick comes from working as many DX stations as does the high-power group, in competition with them. It takes more operating skill with low power, but it all adds up to more fun. They put their emphasis on using efficient antennas. Some hams have worked almost 300 countries with less than 1 W of power. That's a fact.

Do you need to rewire your house for a ham station? Absolutely not. Under receiving conditions (when you are just listening), the power drain is less than that of a reading bulb. Under transmit, the typical set takes 300 to 400 W from the lines, which is not much at all since most wiring is designed for 2000 W, minimum. Not unless you go the full legal limit do you need to worry about special wiring, and then only for that high-powered amplifier (power booster). In fact, I use an amplifier at the legal limit of 1000 W, but I just plug it into the wall outlet. No problem.

Let me tell you something else about this modern gear (going back less than 10 years or so). They are solid. They go on and on, year after year, with almost no maintenance. They built them well, for sure. The only times I have ever had my set out of the cabinet were to make a couple of modifications and once to replace those power output tubes. That's going on 7 years now. (One of those modifications was published in *QST*.)

To summarize what I've said thus far, the basic equipment a novice needs is (1) a transmitter and receiver (or transceiver); (2) an antenna such as a wire antenna of the type we call a *dipole* (see glossary); and (3) a code key. Take your choice: You can do all this for $100 (boat anchors); $300 (used transceiver); $800 (new, moderately priced transceiver); and up.

Stay low to begin. That's what my son did. We got him boat-anchor separates, and he used a wire antenna that was strung in the attic. He worked WAS and DXCC with it. I remember that wire antenna—every time he keyed the transmitter, the bathroom fluorescent light would come on. Strong radio signals can do that to fluorescent lights.

Jim had fun though. So could you. He learned to do with what he had, and he substituted operating skill for fancy equipment. So could you. The moral? Why start with the best? Tough it out with some old stuff so that you get to learn and better appreciate new equipment when you do get it. That way, they'll never call you an "appliance operator"—that is, someone who can do little more than twist knobs.

And, as I always say, your experiences with those cantankerous old sets serve to store up memories for a distant time when you can look at your own daughter's or son's ham radio station with it's flashy new equipment and say, "Back in my time. . . ."

# · 21 ·

# Roll Your Own

There is a special way for beginners to get on the air, and it is a way I favor—build parts of your own station!

Of course, your first reaction is that it is a ridiculous idea. You don't know anything about radio. Well, I've got a surprise for you. Many of those who have said the same thing ended up building some of their equipment.

True, what you build won't be as sophisticated as the commercial equipment. In fact, I quit trying to build anything like the modern sets a long time ago. But whatever you build will have a priceless ingredient that no commercial set could ever have—it will have part of you built into it.

Imagine, if you will, struggling through the construction of a basic and uncomplicated code sender. Imagine, further, that you have it on the bench for a checkout. Say it works fine. Now you are ready for an actual on-the-air contact with it.

This idea of going on the air with something you built with your own hands can't be fully appreciated by newcomers to the hobby until it is actually done. You see, when you build something that you put on the air to make a contact with someone hundreds of miles away, you are joining amateur radio in its finest spirit—the same spirit by which those who came before us had, those pioneers of ham radio communications who always built the strange contraptions by which they too talked to others hundreds of miles away. That's the only way it was back then.

When we "roll our own"—that is, build our own gear (we also call it "home brew"), we have something special. Nothing compares with the proud moment when you casually mention during a contact that "the rig here is a home brew." That makes them all stand up and take notice. You get the respect of other hams right away, and you soon find you are getting questions about your radio.

Novices who build their own code sets are the pride of the band. They may not be able to scoot around as readily as those with commercial gear, nor might they have as much power output or other goodies, but I'll tell you something about that set you build—later, when it has been replaced by a shiny, commercial set, even though that old home-brew job had been laid or mislaid in the garage or attic and become rusted, dented or broken, the memory of building and using it in your early days on the air will be a treasure you will be able to brag about for a long time. Sometimes, though experiences in new things may be keen, memories are what we fall back on to carry us when we are fresh out of experiences. Think about it.

I have another bonus to add to the home-brew approach, and it is this: If you and the transmitter or whatever you build are reasonably normal, the gadget won't work the first time you try it out. That's good. Don't be discouraged. There is no better way in the world to learn about radio communications than by trying to get a gadget you built to work.

You could call someone in to fix it, I'm sure. But if you work over the diagrams and check things, check the manuals and the theory of the circuits and ask yourself what part or parts might cause the symptoms you have, you could figure it out for yourself. And when you do, you will have become a graduate of the self-taught, never-to-be-forgotten school of hard knocks. I did it myself. A lot of us have.

In my younger career I had been rushed through a radio school and became a shipboard radio operator. After the war, I got the ham ticket, and that's when I started building my own equipment. That's also when I began to learn something about radio; a lot, because nothing I built ever worked right. In a way I was lucky for it. I had problems which I recollected when I took my extra class exam 30 years later, and I answered those questions correctly.

Almost all hams own a copy of the ARRL's *Radio Amateur Handbook*. It is an annual bible on amateur radio's many technical sides. I suggest everyone get one as early as possible in their amateur radio career. It even has information on novice studies, although it is best known for its collection of home-brew items for the hobbyist.

Another ARRL publication is the *Weekend Projects Manual,* which describes several basic transmitters, receivers, and many small construction projects you might get into. We all use several small gadgets around the ham shack.

Let's be practical about building transmitters or receivers. They definitely won't be as flexible as the commercial ones. You won't be able to move up and down the band with home-brew novice rigs either. What I'm saying is, you may have to develop more skill and use more patience with a home brew than with a commercial rig. So what? I like to think it's better not to fill up on appetizers but to wait for the main course to come your way, as it surely will one day.

And even if you do have the money to buy a used or new commercial radio, you will still want to learn how to use a soldering iron for connecting circuits. One of the best ways to do so is to build some of the gadgets we all use around the ham shack such as an electronic code keyer. Each such unit you build will require you to develop a basic knowledge of electronics. It will also offer an improvement in your ham skills plus the immense satisfaction of rolling your own, and thereby learning a new skill.

OK, so maybe you won't accept the idea you might one day be building electronic stuff. But if you are a radio amateur, there is everybody's favorite roll-your-own sport of putting up wire antennas. Somehow or other, whether it is because it just gets everyone outdoors or whether it is the climbing thrills we face or whether it is in the fact that anyone and everyone gets to be an expert on antennas without really trying, all these seem to make antennas everyone's favorite when it comes to ham radio equipment.

Wire antennas come in dozens of sizes and patterns, all inexpensive. For a few dollars you get tons of pleasure and use from them. Having so many different kinds of wire antennas is where part of the fun is; it gets everyone talking about "my inverted vee" or "the sloper I use" or "my quad" and other terms you will soon get used to.

Each of the different types of antennas adds their own spice to the soup, so to speak. If you need something short for a short backyard, we've got it. If you need something practically invisible because of a grumpy landlord who doesn't allow outside antennas, we've got it. In fact we've got books written on "invisible" antennas, so if you live in an apartment, don't worry; we'll take good care of you. There is no antenna problem you can think of that some other ham has not had and has solved with typical amateur ingenuity.

Of course, if you are a wheat farmer in Kansas, you have the acreage to put up the premier antenna of them all, the one they call the *rhombic*. Big rhombics run thousands of feet on a leg, and they are absolutely the tops. I have always dreamt of having one.

Whether you feed your radio set into a rhombic antenna or into your aluminum rain gutters (it's been done by several hams!), there are all types of antennas around that you can read about in the many books available on the subject. Antennas are such a popular subject that some of the ham radio magazines put out an entire annual issue on them, and all because they are so popular, so necessary, so easy to build and fun to install.

You will get your first knowledge in novice studies of how long to cut antennas for the various bands, and you will put that knowledge to work when you go on the air. Amateur antennas are cut for the frequencies we use—they just aren't any piece of wire strung between two trees. It's easy, really. Just to show you how easy, let's figure out how long the antenna should be for the 40-m band at, say, 7.1 MHz. All you do is divide the frequency in megahertz (MHz), 7.1 in this case, into the number 462 (you will learn where this number comes from in novice studies). The number 462 divided by 7.1 gives 65, which is how many feet an antenna should be cut for operation on 7.1 MHz. To do it one more time, say you want to put up an antenna on 21.0 MHz, the 15-m band. You divide the frequency 21 into the magic number of 462, and you get 22 ft. As you can see, the higher in frequency we go, the shorter the antenna becomes. That's because the wavelength gets shorter as the frequency gets higher. Remember?

Speaking of short antennas, my son Jim, WA2JNN, used an indoor antenna for a long time and worked quite a bit of DX with it. Many others who live in apartments use indoor antennas, so don't worry about it if you don't have the acreage for a big, long-wire antenna.

As you can imagine, I like to talk about antennas. Any ham does, and so will you one day. But there are also other little gadgets we have around the shack that are fun to build too, and just as easy as an antenna. A lot of satisfaction comes from using something on the air that you put together yourself. Sooner or later you will build something of your own whether you think so now or not. It comes with the territory, this itch to be a tinkerer.

By now you have learned quite a bit about subjects that are covered in the novice exam, and you have learned all this without really trying. For example, you just learned how to figure out how long to make an amateur antenna for any band— you divide whatever number the frequency is (in MHz) into the magic number of 462, and that's all there is to it. You are very likely to get that question on the novice exam, and now, as you know, the answer is going to be easy for you.

Not hard at all, is it, this business about antennas? Nor is the rest of what you need to know to become a licensed radio amateur.

# · 22 ·

# Amateur Radio Magazines

Have you ever gone to the bother of searching all over for information on a new subject, only to later find out that if you had a copy of one of the trade journals on the subject you would have found exactly what you wanted right there? Well, save yourself the trouble in amateur radio because this chapter is a quick look at our national amateur publications.

The comments I give on these publications are strictly my personal opinions and are not necessarily the same views held by others or even the policies of the publications mentioned.

First, of course, is *QST*, the official publication of the American Radio Relay League. *QST* is our oldest hobby magazine and has the largest circulation. Subscription to it is included with membership in the ARRL.

ARRL had been functioning, in 1915, as an organization to relay messages by its participating radio amateurs. But radio amateurs Hiram Percy Maxim and Clarence D. Tuska agreed something more was needed to hold members together.

They decided to establish an official publication, which

they called *QST* (an abbreviation meaning to "broadcast to all stations").

*QST*'s very first issue was published in December 1915. It had 24 pages of news, listings of new stations, advertising and application blanks for joining ARRL. Except for a 2-year period during World War I, *QST* has been issued every month since its first issue.

Today each issue of *QST* covers a wide range of subjects in its features and articles, with up to 200 pages in some issues. When this magazine arrives in the mail, the dedicated radio amateur who reads it from cover to cover (as I usually do) has a few days of reading ahead. I think there is no better way to become impressed with the hobby than to go through each and every section of an entire *QST* issue—I've done exactly that for about 40 issues in researching some of the material in this book. It was an eye-blearying but rewarding activity.

What I also realized from my 40-issue scan was exactly how much the hobby changes and the amount of specialization that is now available. There are over 30 departments written up in *QST* covering clubs, conventions, hamfests, DX, training, operating, satellite operation, QSLs, YLs, and more. Added to this are the feature articles on technical subjects. *QST* has something for everyone.

They also have ads. My, they have ads. You get to drool over the fancy new equipment pictures in their ads. It's enough to make you break loose with the wallet. One issue contained 81 display ads for ham radio dealers or manufacturers. Then there are the classified ham ads in back, of which there are often several hundred per issue, from people who want to sell their used radio sets. Look for *QST* at your magazine stand, for an idea of what I'm talking about.

*CQ Magazine* is a well-rounded publication with a collection of general-interest articles on contests, awards, operating, learning, and building. It's technical level includes novice class, but it is well balanced for all classes of amateurs. It is aimed at the amateur who is active and does things with his or her set. For example, *CQ* sponsors the famous Worldwide DX Contest, held each fall. This is the premier DX contest which spurs us to get our equipment tuned up and our antennas peaked for best results.

*73 Magazine* contains probably the largest number and variety of construction articles of all the magazines. Several computer-oriented articles are also included, making this magazine particularly appealing to those hams who combine hobbies. As you will read in Chap. 27 by *73* editor/publisher Wayne Green, W2NSD/1, computers are where the action is in this modern age, and you will find out that he is certainly the one to know. Typical *73* issues run about 150 pages, with several construction articles on equipment that appeal to the novice as well as the advanced amateur. I especially like their feature in which private radio amateurs from 30 to 35 countries write columns on ham activities in their own countries. And Wayne Green's lengthy editorials are among my priority readings too.

*Ham Radio Magazine* is aimed more at the experienced radio amateur who wants much more in the way of technical material. I have enjoyed many past articles from this publication and more than once clipped one out for some future date.

*WorldRadio* is a monthly tabloid somewhat different in format than the others. It is a newsy, hometown sort of newspaper yet international too. Their format seems to say, "These are the people in ham radio, and this is what's happening." While it also includes technical articles, it is directed more toward news items of public service, emergency support, stories of human interest, personalities in amateur radio, and works of international friendship.

I've written articles for most of the ham publications: *QST* published an equipment modification item I wrote after modifying my radio to make it sharper in tuning; *CQ* has published several I've written, some of which were teaching articles such as one that explained all the radio bands from the lowest to the highest (my diagram made the cover!). One of my most famous articles was my "Bunny Broadcaster" article on our local repeater at the Playboy Casino-Hotel in Atlantic City. *73 Magazine* published the boat-anchor mods I had made to a novice radio, the type I mentioned in Chap. 20 on equipment costs, and an article on tornadoes which is a hobby subject for me. (I've also written a book about tornadoes in the same sort of style as this book.)

Another popular amateur radio publication is the *W5YI Report,* a biweekly, 10-page newsletter that gives its subscribers all the latest happenings in the hobby. With changes as rapid as they are these days the *W5YI Report* is our best way to stay on top of things.

And, of course, there are hundreds of fine monthly newsletters put out by various clubs and groups around the country. A few I receive are the *Amateur Radio News Service; QCWA News* (Quarter Century Wireless Association—members must be in amateur radio 25 years); *auto-call;* and *Collector & Emitter.* One of QCWA's outstanding members, Art Ericson, W1NF, was mentioned in Chap. 3 as probably being in amateur radio as long as anyone else in the world. About 50 clubs feed their inputs to the *auto-call,* plus articles of general interest. *Collector & Emitter* represents about 20 clubs in the Central Oklahoma area (it provides not only club news but several other features including my favorite amateur radio fiction character, the legendary Q.R. Zedd). There are several such fine amateur radio club publications around the country. In fact, the Amateur Radio News Service is an organization of editors and publishers of newsletters who assist editors in improving their newsletters.

So it goes with amateur publications, big ones and little ones. Newcomers to the hobby usually take out an ARRL membership (and *QST* subscription), which I recommend to

all. I also recommend that, as your interests in the hobby develop, you subscribe to one or more of our other publications as the best way to increase your rate of learning in this wide-ranging hobby of amateur radio. You can't find an active and interested ham who doesn't have one or more of these publications coming in. Most of us do; it keeps us in touch.

# · 23 ·
# How to Join

Getting into amateur radio is a process that is a lot like getting into any other hobby—you meet people who are already in it, talk to them, and get a feel for how you like it and how you might do. Of course, this book will serve to answer the question of how you would like amateur radio.

There is one thing about joining amateur radio that I want to make very clear right at the beginning of this chapter, and it is this: We—the radio amateurs, the Federal Communications Commission, and the American Radio Relay League—are all anxious to have you join us. I can't think of a single other hobby where so many people are so willing to give others a helping hand to share their fun. Ham radio is like that.

Many new radio amateurs say they had their start through a friend in the hobby, and that's a great way because you always have someone checking upon you, encouraging you to continue with it. Nothing beats the personal contact for getting you started. Later I'll go into this idea of having a friend help you, but we're so well organized at helping newcomers that I want to go into the formal way first.

Scattered all over the country are amateur radio clubs and instructors who, in the true spirit of developing the hobby, regularly offer courses in novice class studies. They teach these courses anywhere they can find space—their club meeting places or, as is often the case, at local high schools or community colleges. You will read in the next chapter about one dedicated radio amateur who takes time from his busy life to do what he can to bring members into amateur radio. His story is repeated in many parts of the country. Reading about what he has done may show you that we do have such people around to help you join.

How do you get started? Relax. We have it all worked out for you. Write to the ARRL (Amateur Radio Relay League, 225 Main Street, Newington, CT 06111). Tell them you are interested in joining and that you would like the name of the instructor nearest to you. They will match your ZIP code with those on their list of thousands of instructors and will forward this information to you.

No matter what else you do, please do write the ARRL with this request; it gives you a name in your area which can lead to more contacts. ARRL gets over 100 requests like this every day.

The ARRL will send you the name and address of the instructor nearest you. The ARRL will also notify the instructor that you had inquired but that person will probably not do anything about it until you contact her or him. They don't want to be pushy about it. They want to let you make the first move on your own as a way of showing that you do have the interest and desire to join. Some instructors may contact you, but generally they will wait for you to make the first contact. You will get only their address, not their phone number; therefore, your first contact with the instructor will be by mail, even if it turns out that the instructor lives only a block away from you. After that, the instructor might give you his or her phone number.

You can then talk to the instructor to find out when his or her next regular novice class will be held, and where, plus the costs, if any. Most clubs now make a small charge for two reasons—to cover their expenses and to have you make a small dollar commitment in hopes it might encourage you to protect that investment by staying with the course.

With the reply you will receive from the ARRL, you will

also get a copy of their publications, "Answers to Your Questions on Amateur Radio" and a nice write-up on the code, "Morse Decoded." You should read through both of these, particularly the one on the code. It has a subtitle of "Get off and Running Toward Your Amateur License with These Simple Steps to Master the Morse Code." I found this to be an excellent article for beginners.

Usually that's all there is to it. You will have the name of an instructor and literature, including a list of the ARRL publications.

But nothing is perfect, of course, and the ARRL instructor list, being a year or two old, may have names on it of people who are no longer teaching novice courses. No problem. Just ask the person whose name you received if he or she (1) will steer you to someone locally who might be giving novice courses; (2) has the time to work with you, should you decide to study on your own (which is a good approach too, as I'll explain shortly); or (3) can give you the name of a nearby radio amateur who might help you. Be sure to keep in mind my earlier words that most hams are dedicated people who like to bring newcomers into the hobby. If you show a genuine interest and desire, most of us will go the limit in taking a personal interest in you. I find it to be that way, again and again.

In case the name and address of the instructor turns out to be someone who is no longer in the area or for some reason does not reply to your letter, write to the ARRL again to let them know this. They will send you another name, perhaps of someone farther away but still someone you can write to for information on whether anyone in your area knows about novice classes or can help you get started. The important point is that you do not give up if things don't go right the first time.

You will read much more about typical novice classes in the next chapter. For now, I can say that most classes have about 15 students, although I've heard of classes of as few as 2 (one of which I taught myself in our kitchen, and both students became licensed amateurs) or as many as 30 students.

Classes in the novice course usually run one night a week for about 10 weeks. The ARRL has a programmed course outline for instructors which is set up on a 10-week schedule.

You don't lose the whole bundle of wax if you miss one class. In fact, your home studies can be every bit as effective as the classroom teaching, so if you miss one or two weeks, just stay on top of things by studying a bit more at home. Also, if you stay in touch, there is likely to be someone available to help with your questions on radio theory.

In our part of the country, near Atlantic City, we have an instructor who has been teaching novice courses for years. He probably brought several hundred newcomers into the hobby through his courses. In fact, my son Jim, WA2JNN, is one of his graduates. You will read about Jim's developments in amateur radio in Chap. 26.

The percentage of successful graduates from novice classes is very, very high. In fact, you can't miss out on the license if you don't miss out on too many of the lectures. There is no way of measuring this, but my idea is that anyone who has the interest and desire is just about guaranteed to get the novice license.

Now let us suppose for some reason the ARRL route doesn't work for you. Perhaps the instructor has moved or the ARRL list has not been updated for a new name in your area. What to do? You need to find a friend in the hobby if for no other reason than to give you the exam. How to find a local contact? There are several ways.

Call your local high school and community college to ask if they hold amateur radio classes. And if they do not, then ask if they have held any in the past and who the instructor was. That might turn out to be your contact. Some instructors may hold amateur radio courses without having registered their services with the ARRL, so inquiring at your local schools might be a very effective move.

Another way is to look in the phone directory under "amateur radio," where some clubs list themselves. If they do, it will likely be a club station at which there may be no one to answer the phone until meeting nights. If you don't get an answer, keep calling.

And another way is to check the stores in your area that list their products under the heading of "amateur radio." They will likely have radio amateurs working for them. In fact, many amateur radio stores sponsor novice classes. Why not? Any newcomers they get into the hobby are potentially future customers. Although Radio Shack is no longer in the amateur radio market, I sometimes find radio amateurs on their sales staff. Anywhere computers are sold, for example, there might also be a radio amateur salesperson. I bought my computer at a department store where, through conversation, I learned that their section manager was a radio amateur. It happens all the time, this business of finding radio amateurs all over the place. Just keep asking.

Still another way of finding someone is to check out your local newspaper classified ads. We radio amateurs are always selling something or other. In many communities we find that placing our ads in the daily newspaper is a good way to get word to the other radio amateurs. True, we list the equipment in club bulletins, but, for example, in Atlantic County we have about 200 hams who belong to our two clubs; yet there are over 1000 hams in the county. A newspaper ad reaches many of them. Our classified ads may appear under the "radio-TV" or "CB" headings or whatever heading your newspaper might select. Check out all of them. Call one or two names of those who advertise equipment you think looks like amateur radio gear. Don't worry about it if the person you call says, "No, I am not a radio amateur; can't you tell an ad for a hi-fi speaker when you see one?" No problem; a simple, "thank you, good-bye" gets you out of that situation. Sooner or later you will find a radio amateur that way.

Here is another way: Take a short ad in that same newspaper classified section saying "Amateur radio. Looking for local amateur to help with novice study." Personally, if I saw an ad like that in the paper, I would jump at it. I'd know right away that here is someone with enough interest and desire to go looking for help, and I'd be right there, either to help myself or to fill you in on who the local instructor is, what classes are given, and so forth. A lot of us would do the same. Try an ad like that some weekend. Honestly, we're delighted to find that sort of thing.

Continuing with this idea of finding a local radio amateur to help you into the hobby, there is another way to go about it. You could try searching the streets for a home with the kind of long-wire antenna (65-ft or 130-ft) that says a ham lives inside. Or, many of us have beams these days, and they are a positive indication of a ham's residence. Ham beams, by the way, are mounted horizontally—the beam elements (and there are usually three of them, sometimes 5 or 6) are about 12 ft long and often have short (8- to 10-in long) thicker sections along the elements. (We call them *loading coils* because they let us load the transmitter into the antenna on more than one band without adjustments.) I mention this so you can tell ham beams from CB beams. CB beams are mounted vertically with their elements pointing up and down, which is a sure clue to the difference between ham and CB beam antennas. Ours are horizontal, just like TV antennas, and bigger.

When you have located a house that seems to suggest a ham lives inside, grit your teeth and go knock on the door some evening to ask the question. Do it with tact; you don't want them to think you are a complaining neighbor. Make it very clear right off the bat that you are interested in becoming a radio amateur. Nine times out of ten you will, right there and then, make a new friend. Perhaps a lifelong friend. We light up when someone shows an interest in our hobby.

There is still one more way to find a local radio amateur. Do you recall my talking about our amateur radio call book, the one that has the 450,000 of us listed by name, call sign, and address? From App. 4 in this book, which tells what call sign district your state is in, you could get a copy of the call book and spend an hour or two going through that district listing to find someone in your area.

Amateur radio call books are sold by *American Radio Relay League* ($15.75); *Buckmaster Publishing* ($14.95); and *Radio Amateur Callbook* ($19.95). Expensive, yes, but we need them in the hobby and everyone owns one sooner or later. (In fact, I'd suggest you get the first one that has your new call sign in it, to keep around forever. I wish I still had my old 1947 issue—it would have been quite a keepsake to me now.)

With about 450,000 call signs, names, and addresses, these call books are hefty; but they are quite useful. In fact, *Buckmaster Publishing* offers a version that is indexed by geographic locations ($25.00) which would save a lot of searching for nearby hams by having them all listed in order. The other call books list call signs alphabetically broken down by call sign districts. It may take a little time, but it isn't all that hard to go through a regular call book for locations near you. The other route to take while you are looking for a local amateur is to study at home. We have ways for this too. In fact, many people prefer to study on their own. They have found it to be a perfectly fine way to go about it.

One book that is useful in studying on your own is the ARRL publication, *Tune in the World with Ham Radio* ($8.50). It includes a code practice tape for the novice plus a complete list of all questions and answers in multiple-choice format of the 200 master questions in the novice exam. Probably more copies of this publication are sold to future radio amateurs than of any other such publication. Although this is not, in my opinion, the only book to have if you are self-studying, it does have many merits. Many novice instructors also use it for class-room work. The ARRL single code practice tape is not enough, as far as I am concerned, because it becomes necessary to replay the tape too often in going back over some material—which is distracting. I recommend getting more than one code practice tape (other sources are listed in this chapter). The more tapes you have, the better you will find things.

The ARRL has other publications of interest to you such as the *FCC Rule Book, The ARRL Operating Manual,* and *The Radio Amateur's License Manual* (which includes study for the higher licenses but not the novice class).

Novice license manuals are published or sold by several organizations. You can find these at any of the amateur radio supply stores (a few of which are listed in this chapter), or you can write directly to the publishers. Ask for their complete list of novice material, as they often also sell code tapes.

## NOVICE CLASS MANUALS

Ameco Publishing Corp.
275 Hillside Avenue
Williston Park, NY 11596

American Radio Relay League
225 Main Street
Newington, CT 06111

Century Print Shop
6059 Essex Street
Riverside, CA 92504

CQ Book Shop
76 North Broadway
Hicksville, NY 11801

Ham Radio Bookstore
Greenville, NH 03048

Heath Company
Benton Harbor, MI 49022

*73 Magazine*
Mail Order Department
Peterborough, NH 03458

## AMATEUR RADIO SUPPLY OUTLETS

C-Comm
6115 Fifteenth Avenue, NW
Seattle, WA 98107

Electronic Equipment Bank
516 Mill Street
Vienna, VA 22180

Ham Radio World, Inc.
Oneida County Airport Terminal Building
Oriskany, NY 13424

Harvey Radio
25 West Forty-fifth Street
New York, NY 10036

Radios Unlimited
1760 Easton Avenue
Somerset, NJ 08873

Universal Amateur Radio, Inc.
1280 Aida Drive
Columbus (Reynoldsburg), OH 43068

Heath Company's novice study material ($39.95) is a rather comprehensive manual and probably one of the best for the novice available at this time. They guarantee that if you fail to pass the new FCC novice exam after completing their material and practice exam, they will refund your money. What can you lose? The price also includes two code practice tapes.

It is unfortunate, in my opinion, that there has been such a shortage of manuals for the nontechnical student of novice class amateur radio, manuals by which to learn the fundamentals at home. It is my intention to write such a manual along the lines of my chapter on the novice license, particularly the section on electrical principles. I would enjoy writing such a book for the reader who is not the least bit familiar with radio theory. Certainly, there is a need, particularly for those in whom the book you are now reading stirs an interest and desire to join amateur radio.

## CODE PRACTICE TAPES

For the code practice tapes, some sources have already been mentioned such as: Ameco, ARRL, Heath Company and *73 Magazine*. Others are:

Kantronics
1202 East Twenty-third Street
Lawrence, KS 66044

Lockheed ERC Amateur Radio Club
2814 Empire Avenue
Burbank, CA 91504

Twin Oaks Associates
Route 5, Box 37
Knoxville, IA 50138

Wheeler Applied Research Lab
Post Office Box 3261
City of Industry, CA 91794

WrighTapes
235 East Jackson Street
Lansing, MI 48906

I suggest you write to any of the addresses listed for their code tapes and study manuals, and perhaps write to several of them. Look over their literature for whichever items appeal to you. As I said earlier, there is no single novice text available that fills everyone's needs. I recommend you purchase more than one book since they are all quite inexpensive (except for the Heath course). It always helps to get the views of more than one writer when it comes to understanding a new subject.

The advice of having two sets of study material is particularly good when it comes to code practice tapes. Those single cassette tapes are fine if they are to be used along with your code practice in a conducted novice course, but they are not enough if you are learning on your own without an instructor. I suggest you consider tape suppliers who have more than one tape for the novice. The Lockheed Amateur Radio Club, for example, has a 15-tape set ($25 plus $5 shipping) that takes the student from 0 to 13 wpm. Twin Oaks Associates also offers multiple-tape packages which are quite good.

Here is something entirely new in learning the International Morse Code. If you have read each of the chapters of this book in the order in which they were purposely given, one thing you know for sure is my personal affection for the Morse code. What you do not know, however, is of my continuing interest in relaxation techniques by which to calm the mind for new learning. Well, I have put together these two interests in a professionally prepared cassette tape which uses what is known as "subliminal" technique for learning Morse code.

The dictionary definition of *subliminal* is: "below the level of consciousness, too weak or too small to be felt or noticed." That is exactly what my very unusual tape does. Let me further explain.

You do not hear any code sounds on this tape ("Subliminal Morse Code")—only expert voice guidance on one side, and beautifully relaxing music on the other. But the code is in there, along with hidden instructions to relax and to accept code learning as being easy for you to do. You do not need to concentrate with this tape either. You simply turn it on while resting at home or even while driving. All you will hear is relaxing music. Your subconscious mind will respond to what your conscious mind does not; and learning code will turn out to be no problem for you.

Of course, you also need other practice tapes to go with the subliminal tape, but if you use the subliminal tape first, before you listen to any other code practice, you are likely to be more willing to accept learning code than without this tape.

"Subliminal Morse Code" is available from the author:*

Vince Luciani, K2VJ
Post Office Box 682
Cologne, NJ 08213

If you are interested, please write for a brochure.

One other source for learning the Morse code is a personal computer, if you have access to one. Many organizations now offer software (computer programs) which allow computer owners to play code practice. Many of those suppliers of code tapes I have listed also have software packages for the various home computers.

How long does it take to become a radio amateur? Well, if you enroll in a classroom course, you can look forward to becoming licensed as a novice in 10 weeks. If you study on your own, at home, it's a different ballgame because it all depends on how much time you put into it. I recommend you read your study material at least three times each week, spending an hour at each study session. If you want to get the code in a hurry, you should spend another 3 hours a week at it too. With 6 hours of study each week, the average person is ready for the novice exam in about 2 months, perhaps 3 at the most.

The key to learning on your own is to stay with it by regular study times. If you drop it for a few weeks, you may lose ground. Even putting in 15 minutes at code practice is much better than nothing at all. Try to find a certain time of day

---

*The tapes are strictly the property of the author. The publisher takes no responsibility for the quality or effectiveness of the tapes.

when you are most likely to have the time to relax with the novice material. If you have to give up a TV program or two, so what? What you get from TV may benefit you at the moment, but what you can get from ham radio benefits you forever.

If you are working on a one-to-one basis with a local radio amateur who is serving as your instructor, you will probably act the same as anyone else—you will insist you aren't ready for the exam even after 2 months of studies. This is a human reaction I've witnessed many times, that students never feel they are ready for the amateur radio exam. In fact, they very often are more than ready, more than adequately prepared. Listen to your instructor—if he or she thinks you are ready, don't fight it. Take the exam. It doesn't cost anything to try. And if you do fail, you know better what to expect the next time.

Study with a family member. I can't emphasize this notion too much, of how it helps to have someone in the family work with you toward that amateur license. Share those learning experiences, whether with the spouse, the children, grandparents, in-laws, neighbors, or anyone. Amateur radio is great as a family effort.

Grover Conde, WA7USI, wrote to tell me about his ham family and how they shared learning ham radio. Wife Kay is WB7DNJ; son Wally, age 7, is KA7OMP; son Andy, age 9, is KA7OGQ; and son CQ, age 10, is KA7OGR. (Grover mentioned in his letter that Wally's birthday is July Fourth; Andy's is on Father's Day; and CQ's is on Thanksgiving!)

Another ham radio family you might like to know about is that of Dr. Roy G. Clay, Jr., WB5IG. His family includes wife Lou, N5AYU; Roy, III, WB5HVS; daughter-in-law Sacha, N5DUX; Eugene, WD5HGD; Timothy, KA5DBK; and Robert, KA5DFJ. Dr. Roy wrote to me: "Roy III initially got us started in ham radio after reading a fiction story about a ham operator, and playing with a shortwave receiver. Roy III got his novice license first but his younger brother Eugene became the first general class in the family. I was next. Timothy and Robert received their licenses as a result of taking classes. Their mother Lou did, too, and became licensed at the same time.

Sacha got her ticket one month before she married Roy III and I believe it was just as essential as the marriage license. Roy III and Sacha have 2 children; the oldest, Megan, is nearly 3 and already recognizes call signs and some Morse Code."

Yes, indeed, give the kids a chance to share your life. Youngsters are thrilled at the prospect of going to school with Mom and Dad. It gets to be quite a togetherness trip for them, one that is loaded with future memories. Someday they may be telling their children (your grandchildren) of how they went to radio class with Grandmom and Granddad "way back when."

Don't think the youngsters don't have the potential for ham radio. You will later read about how a few grade schools made out with ham radio in the classroom. One school teaches their kids the Morse code and then teaches them how to spell words using the Morse code. You bet, kids take to code like fish to water.

Here is still another way of getting the amateur license. How about radio instructions at a pleasant summer camp? There are several camps which offer amateur radio courses, some of which include the entire family, not just the kids. Some camps you can write to for information on their amateur radio programs are:

CONNECTICUT:

> Lloyd Albin, N2DMQ
> Two Spencer Place
> Scarsdale, NY 10583

MAINE:

> Richard Krasker
> 95 Woodchester Drive
> Chestnut Hill, MA 02167

MASSACHUSETTS:

> Camp Emerson
> Five Brassie Road
> Eastchester, NY 10707

NEW HAMPSHIRE:

> Camp Cody
> Five Lockwood Circle
> Westport, CT 06880

PENNSYLVANIA:
> YMHA Camps
> 21 Plymouth Street
> Fairfield, NJ 07006

VERMONT:
> John Seeger
> Bridgewater, CT 06752

VIRGINIA:
> C. L. Peters, K4DNJ
> Oak Hill Academy
> Mouth of Wilson, VA 24363

With so many summer camps now offering programs in computers, combining amateur radio into their programs seems to be a natural for progressive camps. Besides, if 1000 camps decide to offer amateur radio courses, that's 1000 more jobs created for younger radio amateurs (such as college kids) to get for the summer. For information on other camps that might offer radio courses, write to:

> American Camping Association
> Bradford Woods
> Martinsville, IN 46151

And there you have it. Several ways by which you can go about joining amateur radio, given to you in ways that probably were never before put together quite like this.

To summarize the key points of this chapter:

## NOVICE CLASSES

1. Write the ARRL for the name of a nearby novice class instructor.

2. Attend class one night per week for 10 weeks.

3. Do your class homework, including code practice.

4. Take the code and theory test on your last night at class.

5. Wait for the FCC to issue to you your very own amateur radio call sign.

## HOME STUDY

1. Find a local radio amateur to help you by:
   a. checking the phone directory for an amateur radio club,
   b. calling local high schools and colleges to ask if they have amateur radio classes,
   c. asking at local amateur radio stores for help,
   d. searching through or using the classified ads of your newspaper,
   e. searching for ham antennas in your area,
   f. going through the call book for the name of a nearby radio amateur.

2. Purchase one or more study and license manuals, plus code tapes.

3. Spend at least 6 hours per week in study and code practice.

4. Work with a local ham for guidance.

5. The local ham will give you the novice test and will grade it on the spot.

6. Wait for the FCC to issue your new call sign.

# · 24 ·

# Successful
# Ham Classes

This chapter is about Alan Kline, WB1FOD, as much as it is about successful ham classes because, with Alan, they are one and the same. Alan is a successful ham class organizer. You ought to read his story as proof that we do have people out there who are sincerely dedicated to helping you become a radio amateur.

Alan first wrote his story about novice classes in *73 Magazine,* and we have been corresponding ever since. I was, and am, impressed with his talents and progress.

In the first 3 years of his ham radio classes, he graduated 70 novices, 40 generals, 5 advanced, and 2 extra class amateur radio operators.

Before organizing his program in the Swampscott, Massachusetts, area, he did a survey of needs. There was an ongoing amateur radio class only 10 mi away, so he wanted to be sure he had enough people interested, locally.

He advertised and sent letters to prospective students and soon found the interest to be greater than he suspected.

With that encouragement, he contacted the director of

education at a local school to arrange for use of a large room to use as a novice classroom with space for 25 students and 3 wheelchairs.

Alan wrote in his *73* article:

At this point, I made my first mistake. I offered to pay for the rooms. We had to charge $25 per student to cover our initial costs. For that, they received the *Tune in the World* package and a guarantee of a Novice license! When the time came to make next year's arrangements for rooms, I had learned that most school departments are more than willing to let any civic group use their facilities.

For our second year's program, the town of Danvers, Massachusetts, civil defense radio unit was the co-sponsor. The superintendent of schools and the school committee were glad to give us the needed four rooms at no charge. We met as part of the adult education program on Thursday nights. Both the school and the CD unit helped the club advertise the classes.

During the second half of the school year, we had planned only to teach the follow-up general class and a small advanced class at the Danvers site. But there were so many calls for a novice class, we quickly organized two new novice classes in both Danvers and Swampscott.

Our current class is part of the very successful adult education program of the town of Marblehead, Massachusetts. It is a coordinated effort between the repeater club, the town and the high school's industrial arts department. One class meets after school and the others on Thursday nights.

Our other class is at the Lenox Hill Nursing Home in Lynn, Massachusetts. It is a private institution with many handicapped young people. The director was a radio operator during the Korean War who immediately saw the value of adding ham radio as a weekly activity.

Our teaching effort for this year will be a similar wintertime project at the Greater Lynn Boys Club. As an urban club, its membership swells in the winter months, and not all young people enjoy sports. These 5th, 6th and 7th graders could turn out to be the future electronic technicians this country needs.

Alan stresses informality at their novice classes.

When writing anything for the students, I want to project that ham radio is a relaxing, informal hobby. To help create this informal atmosphere, I always hand-write the announcements for the students. We always stress that anyone who puts in the effort and learns the required code will get the help they need to get their ticket—no matter how long it takes.

On the first night of their novice course, they show the ARRL movies on amateur radio. Alan stresses the value of having students introduce themselves, and invites them to say why they came to class.

With regard to tuition fees, Alan says,

Even if your club underwrites the costs of the class, it's still important to charge for the class. I've argued this point many times over the air, but I am convinced the $10 or so you charge makes the student feel he has made a commitment. No student has ever questioned our fees, and many have made other donations. When asked what happened to the money, I always explained the HANDI-HAM System, because all the extra funds were used to put the handicapped on the air.

After you organize your first class, no matter how large or small, and the first happy novice calls up to say, "I just got my call sign," you'll know why we don't want to stop organizing classes.

# · 25 ·

# The First Time
# Through

In the first part of this chapter I am going to have a little fun
with you by going through an imagined situation where you
have just received your novice license and are going on the air
for the first time.

When that time comes, you tune up your radio and call a
CQ (call to any station), right? Not quite. You see, a first on-
the-air contact can be a nerve-jarring experience worse than
taking your driver's test (although a lot more fun and far more
satisfying).

Comes the day you sit down to your radio for that first
contact (QSO), you make ready to send out your first CQ, you
just might be surprised at your reactions.

Now, don't go blaming me for priming you if you find you
are so nervous your hand slips off the code key or if your pulse
rate zooms. Don't blame me if you find you suddenly feel a
strange loneliness, like you are the only one on the planet and
you are about to discover radio all by yourself.

Huh-uh, don't blame me. After all, this is your thing, your
moment. No one, anywhere, can do this for you, this first radio

contact. Not this one-time event in anyone's life which has been said to be like a strange encounter of the fourth kind.

Relax. Of course you are nervous. It happens to the best, in this first-QSO situation. It happens to everyone. Relax.

But first, think about this: It is only 100 years in man's place on earth that we earthlings have discovered radio communications. Only 100 years out of the few billion years the planet has been around; and here you are, about to transmit a signal into space, hoping someone on earth hears you. Who knows? Who can say whether your radio signal, as it goes out into space, may not continue traveling a billion light-years to a faraway galaxy? Your first radio transmission could do all this? It sure can.

Stay calm. You're not interested in reaching Alpha Centauri. You just want to reach someone in the next town or the next state.

Well, then, let's imagine you have stumbled your way through sending that first CQ call. You have signed your call over and over again, desperately hoping you get it right just once so that maybe someone out there who is listening will decide to take pity on you and answer your call.

And when you have finished sending that CQ call, when you stand by with your fingers gripped unfeelingly on a pencil you intend to write with if you can keep it from slipping through your sweaty, shaking fingers, when you finally stand by and turn to the receiver, almost hoping no one would answer so that you might put all this off until a more relaxed time, some-

how, you stay with it without becoming unglued. You sit there, staring at the set and wondering what's next.

That's when you find out. That's when a strangely familiar sound blares out. A sound that takes a while to sink into the confused thoughts kicking around in your mind. A sound that becomes more electrifying than a full-scale lightning storm. And *then* it registers: "That's *my* call sign! Someone heard me! Someone is answering me!"

My friends, you are about to communicate by code in radio communications in much the same way as did our wireless pioneers at the turn of the century.

May I suggest you make a conscious effort to burn into your mind the emotions you feel at that instant? It will never happen again to you, this first-time-through experience of a code contact on the amateur radio bands. Take my word for it, this is a moment of truth for which there is no modern equivalent—not climbing vertically up a mountain cliff, not spelunking hundreds of feet below the earth's surface, not scuba diving over an exotic ocean floor, nor even in shooting treacherous rapids. None of these offer quite the same emotional wallop as does your very first ham radio QSO in code. Take my word for it.

What would you predict your own reactions might be in this situation? Most of us ordinary people find that we aren't bothered "much" by it. Oh, we may momentarily forget such basics as whether to place a hand or a foot on the code key; we may forget the entire code alphabet in one second, no matter how many hours and days we had been practicing; we may seem to have never heard instructions on those QSO procedures when our instructor would clearly say, "Now, remember. . . ." I've known people to give a happy, blank, stare to the question, "What's your name?" in this situation.

Still, everyone survives the moment, and so will you. Relax. To make sure you do, it would be a good idea to have a ham friend with you the first time through. Someone who can supply you with guidance, understanding, patience, sympathy, or whatever else it takes to get you back to the third planet from the sun. Here's hoping you and I QSO one day so that you hear me say in code, GL ES 73 DE K2VJ.

Do you wonder what it is we say in those code QSOs? Well, App. 5 has some sample code conversations which are typical of those first few you will be having as a novice.

Now that I've got some kidding out of the way (all designed to make sure you capture a memory of pleasure you will long remember), let me tell you something else about this business of ham radio that you won't read anywhere else, a soul-satisfying experience that is available to all radio amateurs of any class license, particularly those who operate in code—the same Morse code used by those who came before us in the hobby.

I like to talk about using code on the airwaves because it is my favorite way of communicating. I think it has much more romance and mystique to it than voice communication has. What I am about to describe may come off just as well on voice, but not for me—I find this satisfaction to come only when using code.

Picture a cold winter night with temperatures down below freezing. A peek outside from within your comfy-warm radio shack reveals a typical winter night in which an infinite number of stars are twinkling through a crystal-clear atmosphere, filling the night sky. The frozen landscape, covered with deep snow, is illuminated by the glow of a brilliant moon. This, my friends, is the season and the time at which radio is at its exquisite best.

The hour is well past midnight. The house is perfectly quiet; everyone is asleep. All the house lights are out, even in the radio shack, except for the soft, reassuring glow of dial lights which cast faint shadows of imagined figures out over the wall of the room to symbolize your favorite activity of amateur radio. Yes, radio conditions are super at such a time. Is this the time to be working DX? For some, yes, but not for you. Not at this particular moment.

No, this is a moment in which there is an extra special reward awaiting you in return for your investment of time and effort in the amateur radio license. This is a moment that is dear to the heart of every true code operator who enjoys casual conversation. This is a supreme moment within the hobby's limitless joys by which to seek the companionship of another hobby member, one who is also awaiting the coincident crossing of your lives; someone you have never met, someone who has also taken this brief moment from the trials of life by which to momentarily submerge his or her problems. This, you will find, is a moment in time when the distant world seems entirely at peace and when life itself hangs loose; when all seems right with the world such as it extends only to the borders of your very private radio room and within the security of your home and family.

This, my friends, is a moment in which to socialize and to make a new friend on the airwaves through the special process of amateur radio as shared by entirely too few people on this planet; this is the time for an amiable, rambling, 2- or 3-h conversation of no particular importance in subject, only pleasure; this is the time for a conversation over the air that will go beyond the subject of equipment and into your personal lives in a way that no letters could. Late-night winter conditions seem to promote such togetherness in ham radio.

And when the two of you have spent some casual hours in idle conversation, it will be but one more of ham radio's satisfactions. And if fate should ever bring the two of you together, 5 or 10 years from that night, both of you will instantly remember—there will be a renewal of amateur radio's bonding spirit as each of you recalls a past, strictly by accident, chance crossing in your lives at a time when the world was momentarily frozen in peaceful calm. To me, this is what amateur radio is truly about.

I have no idea why the setting for the most enjoyable code conversations I have ever had by ham radio have been on crisp, clear winter nights. It may be that conditions are so good then that signals are heard S9, as we say in the hobby ("loud and clear"). It may be that fewer stations are on, so that the bands are more clear. It may be that everything else is put out of mind at that hour too; I just don't know. What I do know is that those conversations (and I still remember some from 30 years ago) bring on such a warm glow of calm and satisfaction that, when turning off the rig at the late hour, sleep comes quickly and easily, spurred by the appreciation of a new friend with whom many things have been discussed.

Try it. You'll like it. In all the world, only amateur radio offers this special benefit.

## · 26 ·

# The Ham in Our Shack, WA2JNN

Our "shack" is our radio room which, at our house, is my son Jim's bedroom. The "ham" subject of this chapter is Jim, WA2JNN. I'd like to have you know a little about him because his accomplishments also represent those which any youngster in ham radio might reach with a little effort.

Jim was 15 when I first asked him to put in writing his inner views of ham radio. He had been a radio amateur four years by then, having acquired his novice license at age 11, in fifth grade; his general license at age 12; the advanced license at age 13; and the extra class license at age 15 (he flunked the extra class exam first time out but did splendidly well the second time). I like to brag about Jim, in case you wondered. Anyway, here is what he wrote:

> I like DX best about ham radio because, to me, it's the prime reason for being a ham. Others like message handling and stuff, but I like DX'ing. There's nothing more satisfying, sitting there at the rig for three hours at a time, griping

because some new country you're trying to work keeps coming back to the other stations. And then, when they do come back to WA2JNN, finally, there's nothing more satisfying than that moment anywhere, not just in ham radio.

Ham radio's advantages are obvious. Not only things like having emergency communications situations but the sheer fun it gives, for one thing. Beats the heck out of watching TV all day. Gives you something to do, some self-fulfillment. Like, you can forget everything that's going on around you by putting out a CQ DX [general call to any DX station] and having someone from some part of the world to talk to.

Contesting is fun, sure. But I really only like DX'ing. Sometimes I rag chew a bit, and sometimes I even go on phone a bit. But usually not much.

I have worked DXCC and have 235 countries confirmed. "Worked All States" was not too bad. I did it on an indoor wire antenna.

I also have the WAC (Worked All Continents), TAD (worked the Ten American call-sign Districts), am a member of ARRL ever since I got my license, I have the ARRL code proficiency award and the RCC—Rag Chewer's Certificate. These, and a bunch of other operating awards.

The beauty of ham radio is doing what I want and like best. Doing my own thing. My fun is talking to other amateurs from other countries, hearing different dialects and trying to deciper their English.

Six months after Jim wrote these words, he had a once-in-a-lifetime thrill from handling an SOS call from a ship at sea. Seems he had been up far into the night looking for new countries when he heard the electrifying signals coming through faintly in code: "SOS SOS SOS from Vessel Lanakai . . . leaking badly . . . position 127 West 17 North. . . ."

This was at 4:08 a.m., our time. Jim calmly checked the maps, realized the position was in the South Pacific and, acting rather well for a youngster in this situation, he called the Coast Guard Search and Rescue Operations in California. They told him to maintain contact with the Lanakai and to do what he could to determine whether it was a genuine emergency or a hoax.

That was when Jim called me. His, "Dad! There's an SOS on," gave me a terrible flashback to decades earlier when I had sailed the seas as a shipboard radio operator during the war. For one instant, I thought it was the skipper of my ship waking me to send an SOS.

The story had a happy ending. Jim stayed in contact with the Lanakai for about an hour until they reported the crew had patched the leak and they were able to proceed on course. Jim so informed the Coast Guard.

The skipper of the Lanakai, who had sent the code, was not an amateur radio operator. He admitted such to Jim. He promised, however, that on his first chance he would study for the license to be better prepared in case he ever had a "next time." What he was referring to was the fact that his code ability was very limited and, if they had been sinking in a hurry, they might never have sent the information needed for a search and rescue mission. We hope he studied for an amateur license although we may never know because the skipper-operator did not get Jim's call sign right. Otherwise, we think he might have written.

The incident suggests two things about ham radio: One, it pays seagoing skippers to get an amateur radio license because there is always, but always, a radio amateur listening somewhere, sometime. Two, when conditions are poor, as they were that night, code will get through, particularly on low power and a poor antenna, when voice might not.

To get back to my bragging about Jim, what I want to add is that Jim is now attending Penn State University as a computer science major. It was a toss-up as to whether he would go into computers or electronics, but when he got his first computer at Christmas when he was 15, he disappeared into the bedroom with it, and we didn't see him for three days. Computers won out, then and there.

No matter. The point is, his future was determined largely from his ham radio work, and there is no question about that in my mind. He is going to be very good at computers because he has also acquired a feel for *hardware*—equipment, that is. He is going to graduate from college with more than a classroom understanding of computers simply because he has his ham radio background.

Of course, some of my background rubbed off on Jim too.

But we parents sometimes have the talent to overdo things to where we convince the kids we don't know what we're talking about. Is your child sometimes affected by that belief? If so, point him or her to ham radio and let this amazing hobby do the talking for you. Just think, in a couple more years my computer science graduate will be in the job market at a cool $35,000 starting salary, unless he continues for the master's degree and the doctorate he's hinted at.

Heed my words, you parents—this could be one of the most important chapters of the book and the most important nudge toward the future your child may ever get. You bet, ham radio is more than just talking on the air. Jim proved as much several times over.

# Wayne Green
# Meanders

Probably the best-known figure in amateur radio is Senator
Barry Goldwater, K7UGA. (King Hussein, of Jordan, JY1, is
also well known.) But the second best-known radio amateur
in the country (and likely the most controversial) is Wayne
Green, W2NSD/1.

It was quite an honor for me that Wayne agreed to take
time from his busy schedule to provide this book with a profile
of his ham radio career. And what a career it is.

Amateur radio started Wayne to his financial fortune. Yet,
for those who know him, his financial fortune is not his true
fortune. We in amateur radio may not agree with some of
Wayne's philosophies but a number of people do appreciate
his individual contributions (one of which you will soon read
about) to the hobby, especially for youngsters.

This chapter could be another very important one to those
with ambitions. I suggest it be read rather carefully, particularly
its ending.

What follows is a summary digest of what Wayne talked

about on the tape he sent to me, a tape he simply labeled "Wayne Green Meanders."

Radio first hit me in 1936, when I was 13. At Sunday School one day, someone brought in a great big box of radio parts for a friend who wasn't too enthusiastic about it, so he shared them with me.

I took the box of parts home, looked in *Popular Mechanics* where they had a little radio you could build using some of these parts, went down to a local radio store to buy the tube that it needed, and put the little radio together. It worked fine.

Then I got interested in building other radios—lots of rigs, receivers, audio equipment, record players, broadcast radios, and shortwave radios. One of the main things you listen to on shortwave is the radio amateurs, and I got excited to listen to the DX come rolling in.

In 1938 I joined our radio club at school, W2ANU, now extinct. I failed the ham test the first three times because of the Morse code (which was required at 13 wpm then). I would get all shaky when I got there and wouldn't copy very well.

Finally, a friend who was going down to take the exam asked me to come along. I figured while I was there I might as well take the exam, as it didn't cost anything. So I sat down to the code, totally unconcerned, not worried about it, and had no problem even though I had not practiced for it. I got my ham ticket, W2NSD, "Never Say Die," in 1940, shortly before World War II. As a matter of fact, the day World War II started, I was on the air where I first learned about it.

My start in publishing resulted from my interest in amateur radio radioteletype (RTTY). I figured there was a need for the hams who were experimenting in RTTY to communicate with each other. When I went to work for a TV station as TV director, they had a mimeograph machine, so I started *Amateur Radio Frontiers,* my RTTY newsletter. After a while we got to a clientele of 400 to 500, and that kept me busy the next 4 years.

I had also started writing for *CQ Magazine* with a column on RTTY, and that got to be very popular too. In fact, *CQ* later hired me as editor as a result of my doing the newsletter and article which I had enjoyed doing, but I hadn't thought of going into the publishing business professionally.

After 5 years I left *CQ* and went to work for an advertising agency but decided I didn't like that. I liked putting out a ham magazine, one that would be heavy on construction articles. So I started *73 Magazine,* emphasizing construction. It did very well right from the start, considering how I had just barely enough to pay for the first issue.

By 1962, *73 Magazine* was still a one-man operation with me doing everything—editorials, advertising sales, typing subscription stencils, and so forth. By 1963 we had some hams working for us, handling most of this stuff, and that made it better for me.

When "incentive licensing" hit ham radio, amateur radio dropped dead. [*Author's note:* Incentive licensing was a change from the previous ways in which you were either a full radio amateur or not, to the current range of license levels as explained in a prior chapter. Wayne has never supported those changes and believes it could have been done differently. He also believes there should be a no-code license, so you can see he does have varied views which should be considered by thinking radio amateurs whether or not we may agree.]

Most U.S. ham radio dealers went out of business— Hallicrafters, Hammarlund, National, Johnson, Multi-Elmac, Davis, Harvey Wells, Gonset, Central Electronics, Webster, Sideband Engineers—all of them.

Studies in *QST* said that 75 percent of the new hams of the time were teenagers, and that 80 percent of them who got into ham radio went on to high-tech [*high technology,* that is, engineering-science] careers.

You will find that the U.S. has lost over 1 million engineers and technicians, and I feel this has a lot to do with our losing one or another of our major electronic industries. For example, when I go into the radio labs in this country, I see only one or two technicians. But when I go into the labs in Japan, I see 50 or 100 technicians. We can't buy technicians here for any price, we are so short of them.

I think amateur radio is one of the best things in the world that can happen to a youngster. I'm a member of the Long Range Planning Committee of FCC, which has in its membership presidents and vice presidents of major companies in America, and quite a few of them started out as hams.

Back in 1975–76, I began speaking to radio clubs and saying that if you wanted to get rich, to get into computers; it was going to be one of the biggest industries in the United States. I particularly meant microcomputers. That's come to fruit. A lot of people, quite a number of people, who have gotten into it have become multimillionaires, and a lot more will. People who are just now getting into computers have the opportunity of getting rich because of the need for many new computer-related products.

When the first microcomputers were invented and when digital electronics came along, it was a natural for *73* because we had been keeping up-to-date by printing articles on it.

Then we started our magazine, *Microcomputing.* People who used that magazine to advertise products, many of them with just enough money to run a first ad, have got up to millionaire. The same thing is happening with another of our insider magazines, with products for the Apple computer. Another new one we started is *Hot CoCo* for color computers. We'll start another, *Run,* for the Commodore computer owners. Companies making products for these computers generally are small enough that there is no way they can distribute through regular channels. About the only practical way they have of selling their products is through magazines which reach the owners directly. *80 Micro,* according to a recent survey, was selling over $30 million worth of products a month.

I perceived the need for a computer magazine because I figured with the invention of the microcomputer there were going to be a lot of people that wanted to learn about them. I found that when I wanted to learn, there were no magazines out. There was a need, and that's why I started *Byte* in 1975. *Byte* is now the largest of all consumer magazines. [*Author's note:* A recent issue of *Byte* ran over 400 pages.]

Since then we've started several more magazines, and in prospect are new types that never existed before. [*Author's note:* Wayne's current publications and enterprises include *73, Microcomputing, 80 Micro, Desktop Computing, Insider, Hot CoCo, Instant Software, Load 80,* Wayne Green Books, Wayne Green International, and Wayne Green, Inc.]

My magazines have been successful enough that I was

recently able to merge with Computer World and got $60 million out of the merger. And that's nice.

I'm reasonably successful, and one of the reasons is I do think up new deals and go out on a limb to try them. Another is I work to make things happen—80 to 100 hours a week for the last 30 years. I do that because I don't know anything I enjoy more. That doesn't mean I don't get on the air or go sailboating or skin diving or things like that, but most of the time I am working, holding meetings, writing editorials and so forth.

I have some ideas to get amateur radio back into a high-growth mode. Number one is to get amateur radio clubs started in every high school in the U.S., and I'm going to work very hard to make that happen.

There are ham clubs in every school in Japan—every one—and I think we should have no less than that here. If we do, we will have a lot more high-tech people, a lot more engineers and technicians. Maybe then we can get back some of our leadership in consumer electronics. It's worth a try. I don't know of any other good way to do it.

With the $60 million, I'm going to be starting quite a few new businesses. We're already well into that. One of them will be a new type of college that will take high school graduates who are gung-ho for amateur radio or computers and give them a chance to really get the background they need to go to work at high-paying jobs. Half of their college time they're going to be working at on-campus electronic businesses which will earn them about 60 percent of the cost of going to college. By having accelerated courses, I think

we can do the whole 4-year course in 2 years. We're going to have some well-educated and highly motivated people come out of that college.

And I think any one of those graduates who doesn't become a millionaire within five years after graduation will be considered a failure.

Business plans for the school [Boston area] have been fairly well worked out, and we're going to see how fast we can get this going.

There you have it, you youngsters who are (1) looking for a guaranteed education, (2) interested in job security in a well-paying field, and (3) not afraid to work for it. Contact Wayne at Wayne Green, Inc., Peterborough, NH 03458. Tell him K2VJ sent you.

# · 28 ·

# Grade School Hams

You are a fifth-grade teacher with students who get upset when June rolls around, students who drop everything and come running back to school in the middle of August. Is this an elementary school, a dream school, or both?

It could be Montebello Elementary School in Suffern, New York, where instructor Bill Lazzaro, N2CF, had infected his students with the ham radio bug.

Bill installed his own ham set at the school one day. As the children entered his classroom, their attention was focused on a mysterious gray box back in the corner of the room. In his letter to me, Bill wrote:

> Ten-year-olds are curious souls, and they had many questions about it. Finally, after much expectation, the magic moment arrived when I fired up the radio for our first demonstration QSO (contact). That first QSO has echoed throughout the halls of Montebello ever since.
>
> It wasn't long before the kids wanted to know how they could join ham radio, to take part in the fun of talking around the country.

At first, Bill thought they were too young to get a license, but he soon learned otherwise.

Since fifth graders enjoy secret codes and ciphers, I introduced them to the International Morse Code. They loved it. There were many more new experiences to follow with the code. By June, sad faces were seen as my students realized their year of introduction to amateur radio had ended.

The following year, Bill began teaching code and theory for the novice exam. The kids did quite well with the code.

The proficiency exhibited by the majority of my students amazed all who witnessed it. By late Spring the first group of six students—three girls and three boys—had passed the novice exam.

Bill's letter to me described their first club station contact at the school: "As we sent our first CQ [general call to any station], six nervous novices armed with pencils and paper were crowded around the set, anxiously waiting. Suddenly we heard the sweetest sound any ham could hear—our own call sign being sent to us in pure notes! It was music to our ears."

The excitement of the moment quickly turned to mild panic as rusty young minds struggled to copy the call sign being sent to us by the other station.
"I didn't get that." "He's sending too fast." "What's didahdidit?"

Since that first contact, however, their club's log book has filled with contacts that include the Azores, Antarctica, and other distant places. QSL cards regularly come in the mail to Montebello School. Bill's letter continued:

The kids really learn code easily. They think it's fun and they love to send code to each other. The theory is a bit of a problem but it can be learned in small workshop groups.
Not everyone passes the novice test the first time. I told my kids that I flunked the first time, too, and I let them know it doesn't matter how many times they fail, they only have to pass once. That seems to reassure them and puts them more at ease on exam date.

The radio amateur program at Montebello grew like a weed, with older children soon stepping in to help the younger ones learn the code and theory. Children would be heard talking about their ham radio contacts instead of last night's TV shows. And when they talk about those countries they contact, they are learning geography in one of the best ways ever brought to a classroom.

In addition to talking to countries around the world, Bill introduced his students to outer space by setting up a set by which to contact one of our OSCAR (see glossary) satellites. He explained:

> We first established an OSCAR receiving station. We learned what a satellite was, how it stayed up, and we studied the on-board systems.
>
> We actually learned to track its orbits and to predict when it was going to be in our range. Then, using borrowed equipment, we transmitted by code through the satellite and heard our very own signals being relayed back down to earth from OSCAR, 500 miles in space. It was terribly exciting.

Bill Lazzaro later accepted an appointment as general manager of the Radio Amateur Satellite Corporation, known as AMSAT—which stands for "amateur satellites." Of the many who applied for this position, Bill's qualifications came through "loud and clear."

Another grade school teacher who has proved the value of amateur radio in the classroom is Carole Perry, WB2MGP. Carole teaches at Rocco Laurie Intermediate School in Staten Island, New York.

Her course, Introduction to Ham Radio, began as a short-term experiment, but, as Carole told me, "It quickly became an extremely popular class, accomplishing several noteworthy achievements."

In their very first term, 300 students enrolled, with 100 children in grades 6, 7, and 8 getting their novice licenses that year! Carole told me:

> It was my experience that with the proper motivation, slow learners and children with behavior and attitudinal problems did remarkably well and showed improvement in other classes as well. It is the kind of program which lends itself to discussions in technical areas—geography, history, current events—and to personal interactions. Almost everyone was

able to find his own reasons for wanting to get involved in the hobby.

I was taken aback at the contagious enthusiasm that developed. Several teachers and parents sat in on some classes and successfully passed the novice exam as well. Even more remarkable to me was watching the fierce determination that many average and below-average students exhibited.

The philosophy of the course was always to learn the basic rules of radio and of how we conduct ourselves properly on the air so that we may enjoy the hobby and all of its incredible rewards.

Carole later informed me that 121 more children had taken the novice exam from a following course, of which 89 had passed. Five students did so well that they went on to obtain their amateur technician licenses.

The ham radio program at Rocco Laurie Intermediate School 72 has caught on like wildfire; each term brings hundreds more students into the hobby. The exciting part about all this is the fact that many of those who do so well at their ham radio studies are the same students who were failing other courses—until something of the old ham radio magic rubbed off, and these same kids suddenly began to improve in all other areas of study!

Now, with Carole's work with the kids so apparently effective, parents have clamored to join in with their children. Stan Katzman, school principal, has developed an evening program in ham radio for those parents. You can expect any number of ham families in Staten Island now, as a result of Carole Perry's exuberance with the hobby and her unselfish dedication in helping youngsters find something new and entirely exciting in their lives. (School administrators around the country, take note!)

Also in New York (surely there must be grade schools in other parts of the country which promote amateur radio?), Joe Fairclough, WB2JKJ, an English teacher at Junior High School 22 in New York City, wrote to me of his success with ham radio in school.

After several years of using conventional methods of teaching English and finding they simply did not work on the youngsters I'm dealing with, I decided it was time for a change. There had to be a better way.

With the idea of creating interest and excitement, I took the standard English curriculum and revised it around ham radio as follows:

1. Teach the children Morse code in the beginning of the term and get them to a point where they can copy their spelling and vocabulary lessons in code;

2. Use the novice handbook as the class text, diagram its sentences, examine its parts of speech, etc.;

3. Reading assignments from the various ham radio publications.

Obviously, there is a great deal more such as QSL'ing, letter writing, geography, math, speech and so on.

Our program receives no funds from any government agency or even the school; we are totally self-supporting. All of our equipment was purchased from the fund-raising efforts of the students and myself, even down to the postage. It's all done by the kids. It is hard to survive this way but it makes for a great spirit of everyone pulling together. And, besides, hams are great people. Without them, none of this would be possible.

Joe's ham radio work at Junior High School 22 was recognized in the following letter to his students from an ex-radio announcer who wrote:

Recently I learned of the amateur radio operation project you, along with the faculty and friends of Junior High School Number 22, have undertaken. Having read the details, . . . I must tell you how impressed I am with this worthwhile endeavor. I can personally attest to the opportunities which radio work can provide. As you may already know, my first job after putting myself through college was with a small radio station out in Davenport, Iowa, as a sports announcer.

The educational experience this project allows you is one that will help you to make sound judgments when you become tomorrow's decision-makers. Congratulations and keep up the good work! (Signed) *Ronald Reagan.*

# · 29 ·

# Senator Barry Goldwater, K7UGA

Ask anyone in amateur radio, which of our members is the most prominent? It always comes out the same—Senator Barry Goldwater, K7UGA.

Amateur radio has a way of reaching into the lives of people everywhere. Some become prominent and still find time for the hobby along with all their other activities. Certainly, being a U.S. senator qualifies one to the claim of being a busy person.

Senator Goldwater took up the hobby of amateur radio at the early age of 12, back in 1921. Licensing was a different process back then and not like it is today. Even so, talent was needed for the license, plus the two necessary qualities of interest and desire.

Obviously, Senator Goldwater had these qualities when he was first licensed in 1922 as amateur radio station 6BPI; this accomplishment truly qualifies him for a certain prestigious title whose understanding is more recognized in ham radio than in many other areas—Old Timer. Senator Goldwater wrote to me about his earlier years:

> There wasn't much of a way to practice code with other amateurs in those days since there weren't many of them. I would just listen in and find somebody whose speed I could copy and I gradually improved until I got to ten words a minute, which was all we needed at the time.
>
> The exam was extremely simple then, compared to the general class exam today. The only advice I can give to prospective novices is to devote about an hour a day to studying the technicalities and listening to the code. You can get to the point of no return from over-practicing.

When I asked him about his first QSO on the air in 1922, he replied, "I can't remember who was my first QSO but I do remember the extreme pleasure and excitement of it. In fact, when it happened I remember yelling all over the place for the family to come and listen. All I had were earphones and I had to split them up for the others to hear."

> My favorite band is 20-meter phone, although I also work code occasionally on all other bands.
>
> My activity in amateur radio is in just keeping active. Most of the work I've been doing the last 15 years has been phone patching [see glossary] and RTTY ["radioteletype," see glossary] for Air Force members who are on overseas assignments. The most valuable memories I have in ham radio are of the phone patching we brought to servicemen, allowing them to talk directly to their families back home."

What the senator didn't mention was the astounding phone-patching record set by his station—over 200,000 phone patches overseas in an 18-year period. This was with the help of various local radio amateurs who operated from a special station set up for handling personal military traffic by amateur radio. The station was eventually closed in 1983, after many years of admirable service as a fine credit to amateur radio.

Although there is an amateur radio station, W3USS, in the senate building, Senator Goldwater seems to be the only elected member of Congress who is a radio amateur.

K7UGA defines amateur radio as more a public service than a hobby: "I call it a service, and that's the satisfaction I get from it, doing something for somebody else. Ham radio has provided me with a lifetime of pleasure and enjoyment, relaxation and a chance to do something for someone else."

Another point about Senator Goldwater is his consistent participation in amateur radio. He certainly is not a token member, as evidenced by his regular appearances at various amateur radio events.

Amateur radio is rather proud of his presence in the hobby.

# · 30 ·

# Chet Atkins, WA4CZD

For $275 you could buy a new Ford touring car. A loaf of bread was 8 cents, and a pound of lunch ham cost 25 cents. Those were the days.

Those were the days—June 20, 1924—when this nation brought forth a guitar-pickin', country-music-singin' native of east Tennessee—Chet Atkins.

As a youngster of 11 in the small community of Luttrell, about 20 mi from Knoxville, Chet picked up his brother's guitar and hasn't since put it down. Guitars continue to represent one of the most important joys in his life. "Sure beats working," is the way he said it.

Now a giant of the recording industry, Chet and his guitar have continued to "beat working" through such accomplishments as sales of albums in the millions, guesting on national TV, solo performances with symphony orchestras, Grand Ole Opry, and various worldwide tours.

Today Chet Atkins, known to us in the amateur radio hobby as WA4CZD, continues with his recording work out of Nashville. He graciously detoured from his work long enough to provide me with the following story of his amateur radio life:

> My first introduction to ham radio came when I was a teenager living in the mountains of East Tennessee. There was a fellow up in one of the hollows who built radios. He didn't have a ham license at that time but his work on radios stirred my first interest in the hobby.
>
> I later built a transmitter from a diagram in a ham radio magazine. But I didn't get it on the air because it was then wartime.
>
> Later, in the 1960's, I had a teenage nephew in Denver, Gary Atkins, W0CGR, who is a very good ham. When I went to visit him I was attracted to his QSL cards and the nice rig he had.
>
> This, again, stimulated my interest in ham radio. Gary wrote the code out for me, which I rehearsed by reading everything in code that I saw, numbers and so forth. Then I bought a receiver and started listening to the W1AW code practice sessions.
>
> [*Note:* W1AW is the official ARRL broadcast station which regularly sends out code practice and bulletins over the ham bands.]
>
> I never took any classroom training. I went the self-study route all by myself. I was very pleased to have passed the test the first time out. I knew I was OK on the theory but the code had me a bit nervous although, actually, I found that being a musician and having a trained ear helped me quite a bit. So did learning to type by the touch system, which is somewhat like learning the code.
>
> [*Note:* Those of us who had any musical background— I used to play the clarinet—know that what Chet says is true about the code being easier for those who can read music. His comment about typing being a help for code is also true, as many secretaries who are now radio amateurs have learned.]
>
> As far as activities go, I never went in for contests. Mostly I work quite a bit of DX [foreign contacts] and some 80-meter phone [voice contacts] to old friends in East Tennessee. But most of my hobby was, and is, DX. I also work two meters into the repeaters and carry a walkie-talkie with me with a touch-tone pad [for telephone hookup] in case I have a problem.

Sometimes I use the walkie-talkie to rag chew with locals, such as when I'm driving to the airport. I also enjoy talking to Gary in Colorado every Sunday afternoon. I enjoy that very much. We have mutual interests. He's much smarter than I, in electronics. I find I can learn a lot just from talking to him.

I think the amateur hobby is a heck of a nice way to meet interesting people and to have interesting conversations about equipment and electronics. If you're an introverted person like I was and am, it's a good hobby to help a person. I know I used to have trouble talking on the microphone, which ham radio helped improve quite a bit.

A lot of people may be scared away from ham radio by the code and theory requirements but I can only tell them it's not that difficult. The reason they may get scared is that they don't really apply themselves. They should work slowly and regularly to build up their code speed. The electronics for the novice exam is so easy it can be learned quickly. I see no reason why anyone who applies himself can't become informed enough to pass the novice test.

One good way for people to get interested in ham radio is to first get a receiver and to start listening to hams around the country and around the world. That's how I got interested, anyway. I love to listen in. Once I had heard those guys on the air, I immediately decided I wanted to get on myself, one day.

It's a great thrill to make your first DX contact on code. Talking all over the world on code is a terrific thrill.

It's a great hobby and I recommend it for anyone who might have the slightest bit of interest.

Amen to that. In particular, the hobby is ideally suited to entertainers who are on the road a lot. You will read another entertainer's views of ham radio in Chap. 32, when Larry Ferrari adds the glue to Chet's recommendations.

# · 31 ·

# General Curtis Le May, W6EZV

Amateur radio operator Curtis Le May, W6EZV, is the American Air Force general you history buffs may remember. He developed advanced bomber tactics in World War II and earned the nickname of "Old Ironpants" from his daring daylight raids on Berlin through heavy fire.

He will also be remembered for having so well organized the famous Berlin airlift shortly after the war, when that city was cut off from all ground transportation.

Born in Columbus, Ohio, Le May became a fighter pilot in 1929 and became the youngest general in the Air Force at age 37 in 1943. After the war he headed the Strategic Air Command until 1957 and later became the chief of staff of the U.S. Air Force. I've talked to several World War II veterans who remember General Le May, some of whom were on his staff at SAC.

The following profile of General Le May's amateur radio

life was from a phone interview I had with him from where he lives in retirement in sunny W6-land (California).

I first heard about amateur radio in 1935 when I was assigned the job of communications officer of my squadron. I didn't have a license, then. I just got parts and put them together in a ham station for others.

I got my first ham license in Germany, right after the war. Things were in a turmoil, then, and one of the things we had to do was to dispose of all the supply depots scattered around Europe. We had so much surplus radio gear, any ham could go to the signal depot and draw equipment to set up a ham station. My call sign, then, was D4AFE.

When I was back in the States, commanding SAC at Omaha, I found that the MARS [Mars Affiliate Radio Service] was deteriorating, so I made a deal with my deputy commander that if he studied up for the ham license, we'd get on the air. I had to bone up on the code, which I used to be able to handle at 12 words per minute in flag school, and built up my speed to well above the requirements to pass the exam. This was in 1947.

When I started operating on the air, the first things I noticed, those days, was you had trouble competing on phone [voice contacts]. Yet, there was a bunch of guys on the bottom end of the band on single sideband [also voice, see glossary] who seemed to have a ball.

My deputy commander had gone to school with Art Collins [Collins Radio Co.], so we borrowed a single sideband [SSB] set from Art and set it up at my house.

This was much better than the old phone sets, so we decided to put one of the SSB sets in my staff plane. The deputy commander went out to the Far East with it and I worked him clear to Tokyo on the ham bands.

I was convinced that single sideband was much better than what we had in our Air Force planes those days, and I tried to convince our Signal people to look into it.

But they didn't think it could punch through the absorption belt over the northern hemisphere, so just about then I decided they'd better send someone out, pronto, to look into this.

I remember they sent a second lieutenant. We put him on a plane and sent him off to Thule, Greenland, while we worked the plane from Omaha all the way, even when they landed at Thule. We worked them on the ground all night long on SSB.

Next morning they took off for Anchorage over the North Pole. I sat in my house and listened to hams 10 deep working the plane as it crossed the Pole. I recall it took us nearly 3 months to get all the QSL's out because SSB was new then, and a plane over the North Pole was new DX [foreign contact].

From that experience I arranged SSB ham radio stations at Goosebay, Labrador, Azores, Morocco, England, Hawaii and Guam. We manned them 24 hours a day, keeping regular contact every 15 minutes. The SSB equipment turned out to be so much better than the old phone transmitters that we finally got SSB for all Air Force planes. Now, everyone is in the act. You just can't beat SSB.

I'd say my most memorable experience in ham radio was in getting the military to use ham radio with SSB as part of military communications units everywhere. We used ham SSB all the time I was in SAC and we had units all around the world, each equipped with a ham station. It gave them a chance to stay in contact and to run phone patches [connections between radio set and telephone line] back home.

These days, I just get on the air to talk to a half dozen old friends every weekday. I use a simple doublet [antenna] stretched up in the attic. I once did a little radioteletype work but not anymore. I hardly ever had a chance to mix into too many ham radio activities. Mostly, I had fun building things, working with my hands on something. I built a lot of kit units.

There were times I spent in Washington that I tried to keep up with ham radio but there never seemed to be enough time to do as much as I would have liked.

I set up a station when I retired out here in California. At first I didn't do much except listen in. Then I contacted a few old friends, and that's what I'm doing mostly in ham radio these days.

I think ham radio is a hobby for people who like to get into it for something to do, and it's very interesting. You may even end up with something especially useful for others, like what I did in getting SAC to move into SSB and ham radio.

Hams have always made great advances in radio communications.

# · 32 ·

# Larry Ferrari, WA2MKI

From my old home town of Mahanoy City, in the hills of Pennsylvania, back when television was barely getting started in the 1950s and when any of the TV pictures you got were weak and shaky, I have a clear recollection of a program that this new entertainment-in-the-living-room gadget brought to us—of Larry Ferrari doing his own musical magic at the keyboard of the Hammond organ.

Even years later, at age 85, my mother used to remember to turn on the TV every Sunday morning to catch Larry's show. Despite her failing hearing, she liked what she heard. Larry's fantastic handling of the keyboard was and is sheer artistry. It is quite a pleasure then to be able to tell you about the amateur radio life of WA2MKI.

Larry began his television career while stationed at Fort Dix, New Jersey, during the Korean war, with a program called *Fort Dix Presents*. When he was later mustered out, talent was recognized and they asked him to stay on TV with what, after all these years, is still known as the *Larry Ferrari Show*.

While laying claim to no particular electronics background, Larry told me:

> Music is my life, true. But the secondary side of it has always stayed there and that has been my personal interest in all things electronic. Amateur radio finally got to be the means to that end.
>
> My first introduction to amateur radio came while I was stationed at Fort Dix, through an old friend, W2RQC, Bill. Bill had a terrific ham station which I was always in awe of because it all looked so fascinating yet so foreign to me. Bill recognized my interest and explained about things like the code. Unfortunately, I had a lot of other matters on my mind, then, and didn't take any action.

When the CB boom hit, Larry and a group of associates at WPVI-TV, in Philadelphia, had mobile radios by which they would talk to each other on the way to work. But CB served to build up Larry's appetite for really good communications, the kind that only ham radio offers. With the encouragement of other hams at WPVI-TV, he made up his mind to go after the amateur license.

> I didn't take any classes. I just studied on my own and soon acquired a novice ticket. I had the ARRL course material and the Ameco book to study from.
>
> Since I was on the road a lot, both by car and by plane, I took the material with me always. George, who assists me, would send code to me while I'd be driving.
>
> On plane trips I'd simply read the material over and over again, those being my best moments for concentration. Shortly after acquiring the novice license I went for my technician license and promptly followed that with the general class license.
>
> I did it all on my own, although the fellows at the station, about 12 are hams, were a big help. One of the cute things I'd like to tell you about is the way they always knew at the studio I was into ham radio because I would send out practice code on the big Hammond organ. Now that's what you call a really expensive code sender!

While the FCC hasn't approved the Hammond organ as a ham radio rig, Larry admits to starting more than one QSO (contact) with it. For example, he mentioned a tour of Las Vegas where, after a while, a few people would drift up to the stage and begin a conversation with, "I'm W6 so-and-so. Or, I'm W8 so-and-so, and I understand you are also a ham radio operator." Larry delights in such introductions.

It pleases me to have people at a concert somewhere introduce themselves as amateur radio operators. I've gotten to where I can almost immediately tell when someone walks up that they are going to give their call signs. I have made many new friends that way, and they are all really nice people.

A lot of those friendships I still sustain through on-the-air schedules. I can't say too much about how ham radio has added to my life because I so thoroughly enjoy those nice conversations with hams and the friendships they have brought me.

The most important thing with me in amateur radio has been and always will be my natural fascination with all things electronic. Amateur radio is the extension of this interest via worldwide communications with people. My television appearances are, to me, a people-to-people thing which I enjoy, but amateur radio is also a very important and personal matter the way I see it.

I enjoy the friendships and the camaraderie you get from other hams in other parts of the country and the world. That's the whole thing in a nutshell. I thoroughly enjoy meeting people and talking to them on the air.

I've had a lot of nice things happen to me in my ham radio life. For example, there was the time we were driving to Richmond late at night when I fired up the two-meter rig on a local repeater and contacted a W4-station, Mac. In the course of our conversation I explained what we were doing and where we were headed. Mac mentioned we would be driving right by his house and invited us to stop by.

We figured on a five-minute coffee break but we were soon wrapped up in Mac and his XYL's [wife's] hospitality, and Mac's other hobby of woodcarving, that it was an hour before we got back on the road. Typically, that was one of the many nice things about ham radio I have experienced. Picture yourself driving along a strange highway one minute, and the next minute you are in the home of a new friend who treats you like a lifelong acquaintance. Where else but in amateur radio?

Larry's closing comments are of particular value to readers:

> People who are interested in the amateur hobby shouldn't allow themselves to be turned away from it. And you'll never regret it.
>
> Particularly, more people in the entertainment field should take the time to learn what amateur radio is about, and what it can really offer.

Larry is right, as Chet Atkins, Arthur Godfrey, and several other people in the entertainment field have learned. Donny Osmond was a radio amateur too. And I remember doing my best to get Dick Van Dyke into the hobby via his close friend and former producer, Byron Paul, WA6RNG (but I don't think I've yet convinced Van Dyke, which is a shame because he would be a natural to the hobby). Entertainers really can benefit from ham radio more than they might ever suspect.

# · 33 ·

# Reese Palley, KA2AUP

Boardwalk at Park Place, Atlantic City. At this Monopoly-famous location, one finds the unique art shop of entrepreneur, salesman, sailor, author, and self-professed "peddler with a fancy pushcart," Reese Palley, KA2AUP.

This second-generation Atlantic City resident is an internationally known figure whose sales spectaculars have included chartering a pair of 707s by which to fly customers to the opening of his Paris gallery. His stylish salesmanship reportedly causes weeping and gnashing of the teeth at such shops at Bonwit's and Neiman-Marcus, particularly when they encounter his famous slogan, "merchant to the rich."

Reese Palley is a sailor at heart. He now owns a huge sailboat with which he has covered the Orient. How did such an individual find a lead into ham radio? KA2AUP gives this story:

> I'm a sailor and I sail long distances. It is necessary for me to be in touch with my family and home. Normal ship-

166

shore services don't provide the extras I need, including the emergency situation. Ham radio does.

When I contacted Herb Johnson, president of Atlas Company, a while back—and Herb is also a sailor—his prompt advice was for me to get an amateur radio license, which I did.

I like code best. With code you can more easily get through. My original reason for putting ham gear on the boat was because I wanted to be sure of reaching someone in case of emergency. If you put out an SOS on code, you're going to reach someone fast. It gets through, and accurately.

Although I had no particular knack at it, I loved learning the code. I liked the idea of acquiring a whole new skill. What's exciting to me about learning code is, for example, when you first start hearing diddahdidit for "L," and not did-dah-dit-dit. When you see the letter in your head without going through an intermediate step of identifying what it is, that's when I really enjoyed it.

I think the day we have ham radio without code is the day we won't have an amateur service anymore.

I started off with a pair of Atlas radios for my equipment, and later added a Palomar set. Then I realized that if we sail into some place like Dakar, and I'm looking for some local hams—and in a strange place, that's exactly what you want to do—I figured I needed a two-meter set. When I sail, all the gear is on board. I always like to have back-up equipment.

I like ham radio as a hobby in the sense of learning about radio and technical stuff but not strictly as a chatterbox hobby. I like the many nets—the missionary nets and the maritime mobile nets where you can check in from the boat at any time and people keep track of you.

I also like the medical nets. I think every physician should be a ham, to plug people in anywhere in an emergency situation.

Little did Reese suspect that sometime after our interview he, himself, would be doing that very thing—of "plugging" in a doctor (who was his son) into a ham radio call in an emergency situation (his own).

Reese described the incident in which ham radio at best saved his life and at least saved him much pain and possible scarring. They were on his sailing ship *Unlikely VII,* 2000 mi from the nearest landfall in the South Pacific. Two weeks at sea, and having grown a bit careless, Reese fired up their alcohol-primed kerosene cook stove except, on that morning, he used entirely too much alcohol to prime it. Flame shot from the stove and up to the bottle of alcohol still in his hand, which he threw aside when it ignited, causing more fire to spread around the galley and about his person.

Second-degree burns covered his body. Reese's first thoughts were of his physician-son, Gil. Through a U.S. radio amateur who patched Gil to Reese via the ham radio bands, Gil explained that he had packed his dad's medicine kit for the trip with huge amounts of burn ointment.

It was fortunate for Reese that Gil had done so. In the 2 weeks before they reached land, Reese followed his son's advice, particularly with the ointment. By landfall, Reese reported no more than pink skin areas where the burns had been. Had he been stocked with what he previously thought to be a "normal" supply of burn ointment, he suspects he might have been in quite a bit more serious condition, perhaps with permanent scars.

And having his son-doctor available via the ham set in the days they sailed for port was, I am sure, equally good medicine for Reese.

# Rosemarie Deisinger, W9AWI

Louisa Sando, W5RZJ, wrote a fascinating book called *CQ YL* in which one of its interesting chapters on the women in amateur radio was a reprint from *QST Magazine*—the story of Rosemarie Deisinger, W9AWI, a blind radio amateur. Here is the story as told by Rosemarie:

My introduction to amateur radio came in the Spring of 1947. Until then, the word "ham" signified nothing more to me than a picnic delicacy with which one served German potato salad and cole slaw.

That Spring, I came to know Johnny Ludwig, a bedridden veteran with multiple sclerosis who lived in my parish. Johnny liked to do nice things for people and he sent me a subscription to a Braille magazine. When I went to thank him for it, I found Johnny was not only a wonderful person but a ham, W9BLY, as well. His sickroom was cluttered with equipment, all strange to me.

Moreover, I learned that Father Andy, our priest, was just then in the process of preparing for his license exam. He took the test soon afterward and got his call of W9ZNE. In my succeeding visits with Johnny, he began to drop broad hints to the effect that Milwaukee needed a YL [young lady, ham term commonly used] operator, a rare species of humanity missing since the wartime clampdown on amateur radio. Usually, I ignored those innuendos but one day Fr. Andy tricked me into admitting that it might be interesting to learn about hamming. With that innocent remark my fate was sealed. A license manual, a code machine and records all arrived at my house within a few days. Well, the challenge appealed to me so I plunged in.

The code came quickly and easily. My good sister-in-law helped me master the manual, mainly as a matter of memory. Fr. Redig helped me with the diagrams, and Johnny taught me theory.

To my horror, Fr. Andy announced one June day that he had made an appointment for me to take the exam in Chicago. The examination went very well, and on June 21, 1947, I received the call W9AWI.

The good Father had a rig all set up in my home and, by prearrangement with Johnny's mother, his receiver was set to my frequency. When he heard me calling him, he was too excited to work the controls. It was a very happy occasion for both of us.

That same evening, I contacted a W9KSP, another Johnny, whose "handsome voice" I had often heard while eavesdropping, before I got my call. It turned out that he was also a good friend of Johnny Ludwig, W9BLY, and the three of us had many wonderful QSO's together.

Johnny Ludwig's strength faded, and in the Fall he died. I think it is a great tribute to the hobby he loved so well that, in his dying hours when he was delirious, he signed off to all his old ham friends. You can understand with what fondness and admiration I will remember him.

In August, 1950, W9KSP—the "handsome voice"—and I were married.

It would be impossible to measure the enjoyment, happiness and thrills hamming brought me that summer and in the years since.

Blind people are severely handicapped in the ability to get out and make friends. Amateur radio brings a host of friends from all over the country right into the ham shack— and they're real friends.

I would certainly say to an aspiring YL who is blind that the rewards justify any expenditure of time and energy.

Such was Rosemarie's striking story.

Now, after more than 35 years, I wondered what had happened to her. Rosemarie wrote to me:

> Your letter came as a complete surprise, and with Christmas just three days off. I promise you a letter after the holidays when the smell of gingerbread has cleared the air, when I have brushed most of the dried Christmas tree needles off the windowsills and extracted most of the nutshells from the carpeting under the sofa.

She later wrote, "For one thing, there have been eight children to bless our lives—Regina, Francis, Thomas, Ann, Christopher, Rita, Mark and Rebecca."

For the first 13 years of marriage they lived with her parents. "So I had plenty of help with the babies who all had to learn to talk early so they could communicate with me. Just pointing at something they wanted was not enough. They had to talk."

Later, home became a beautiful house in Wauwatosa, Wisconsin, where Rosemarie still finds time for her two favorite hobbies of classical music and amateur radio.

> I have had several memories from those happy years of hamming. One was a chance QSO [contact] with a mobile station in Guatemala City. When I told him my sister-in-law was a nun assigned to a convent there, he drove directly to the building, located that astonished nun and took her out to his car where we had a marvelous talk.
>
> My most thrilling experience in amateur radio came when a station broke into one of my contacts. It was a UA3 [Russian] station calling me. I really blew my cool. I had never even heard a Russian station before and couldn't believe I was being heard in Moscow!

Rosemarie continues to operate the amateur radio bands, particularly to talk to daughter Regina who volunteered her services as a registered nurse at a charity hospital in the Caribbean island of Santa Lucia. Although Regina is not a radio amateur, the hospital's administrator, Sister Mark, VP2LBR, is. Rosemarie added:

> I would like to work DX [foreign countries] again. Not to chalk up any records but simply because I like people whether they live across the street or across the world. When I contact a station I am always looking for a thread in the conversation I can snatch out and weave into the sampler hanging on the "wall" of my heart . . . *People Count!* And hams are very wonderful people.

# · 35 ·

# To Quote a Few

Handicapped radio amateurs make use of many different special gadgets to help them operate their amateur radio sets. There is, for example, a *tactile* (sense of touch) converter for deaf hams, a puff-and-sip code keyer for quadraplegic hams, tone devices for the blind, and a host of other items, all of which are available to make sure that no matter the physical handicap, amateur radio's handicapped operators have ways of getting on the air by themselves.

Do they think of everything? Probably not, but they do have a case of universal allergy in which helpful hams have come up with an airtight, stainless steel, enclosed station. And they also have gadgets that convert transmitter dial settings into voice announcements. There's no end to what they can do these days.

Here are a few short stories of handicapped hams who make out quite nicely, thank you, on the ham radio airwaves using a few gadgets to make up for their handicaps.

Gayle Sabonaitis, WA1OPN, is a blind and deaf woman with real talents who managed to open the entire world to her touch simply by learning about ham radio. Gayle operates by code, of course, but not at any slowpoke speed, I can tell you. She zips right along at better than 20 wpm.

How does she do this so well? She receives the code through a gadget connected to her radio receiver which changes the code sounds from the loudspeaker into vibrations. By placing her fingers on the vibrating surface, she can pick off those vibrations just as neatly as can be. Stop and think about it— 20 wpm averages out to about 100 code characters per minute or almost two different letters every second that she "hears" with her fingertips.

So now you know one of my secrets. In my talks on amateur radio, I always like to have the audience challenge me when I talk about how the deaf can "hear" in amateur radio and of how the mute can "talk." By code, of course.

Gayle is so good at the hobby she is an extra class radio amateur, for which she had to pass a code test at 20 wpm. Terrific woman, that Gayle. President Reagan agreed. He sent her a personal letter which commended her for her work in helping disabled people obtain amateur radio licenses. Amateur radio is quite proud of Gayle.

Dick Eichhorn, KB0AE, is also a blind ham. He is active on the air with one of his favorite activities—a gathering of a whole bunch of hams, most of whom are handicapped, in a group they call the PICO-NET. (PICON is a saying in ham radio that comes from the FCC definition of the hobby as being "in the *public interest, convenience or necessity*). If you listen to this group on the air, you have no idea any one of them has any kind of handicap. On the air, they don't, and that's the simple truth. Amateur radio is such an equalizer that the only "handicap" we ever recognize is negative thinking, no matter one's physical nature.

Kay Clark, VE3KAY (what a nice call sign for someone named Kay), lives in Caledonia, Ontario. Both deaf and blind, Kay also uses a device that transforms sound into vibrations in order to copy code.

When Kay first learned ham radio theory, she did so from instruction tapes that were in code! Now there's a double twist for you. She had to be good at code in order to learn theory. Kay was the thirteenth deaf-blind licensed radio amateur in the world. She is quite active on code on the ham bands, does a lot of contacting all over the world, and, in fact, wrote to tell me she has many special ham friends in Japan who are interested in how she operates. And she does all this from an apartment with a small vertical antenna and a little radio.

Mary Hauck, WA6VUE (now deceased), at the fine age of 85, though blind, was sending out code on the ham bands. She had some understanding relatives, by the way—11 of them are hams, including her 3 sons and their wives.

WD6BPT, "Bed Pan Trainers," they call themselves, is the club station at Saint Jude Hospital (not the one made famous by Danny Thomas) in Fullerton, California.

Why a ham radio station at a rehab center? April Moell (somehow I relate to the name of April so well, I instinctively know what a beautiful person this lady is) is radio amateur WA6OPS, who wrote to me that amateur radio at rehab centers could provide "motivation, stimulation and orientation for our patients."

I didn't know what that meant (at least I pretended I didn't), so April came back in my language when she added: "A happy patient heals faster. And there is nothing like the smile on the face of a patient who hears his or her name on the radio." *That,* we understand!

Their patients are victims of stroke, head injury, spinal cord injury, and other disorders requiring them to learn, again, to talk, walk, and feed themselves. Ham radio seems to produce the best from these patients. They outdo themselves at the

microphone, having to relearn to talk, simply because someone on the radio mentions their name. And, by golly, they want to say "Hello, how are you?" right back. That's motivation. Tell me about the quarter-million-dollar machines they have to do the same thing, and I'll tell you about how ham radio does it cheaper, better, warmer, easier, friendlier, etc.

Anyway, April said some patients are helped with visual scanning simply from using a large map on which they pinpoint where the other station is talking from. Stamping and filling out QSL cards also helps with coordination. Others, if they wish, can practice sending and receiving code with a little code set.

Rehab radio, inspired by a special person, WA6OPS, is about a half dozen years old and not only going strong but growing. April travels around her part of the country, responding to numerous inquiries from other facilities wanting to set up amateur radio programs similar to hers at Saint Jude. I don't know for sure, but I have a hunch that if your organization or your patrons were willing to provide travel expenses, April (and I, as a matter of fact) would come to your facility anywhere in the country to fill you with wonderment as to the marvelous improvements in the lives of rehab patients, those who take to ham radio if only to enjoy listening to others operate it.

The Braille Institute (714 North Vermont Avenue, Los Angeles, California 90029-3594) is a private, nonprofit organization serving, without charge, the blind of all ages of southern California. It was founded in 1919 by J. Robert Atkinson, a blind Montana cowboy who emigrated to Los Angeles.

They offer five basic classes for the newly blind and 123 other classes of which ham radio is one. They also work with the HANDI-HAM System (next chapter) in a West Coast camp, at Malibu, in ham radio classes. This chapter is already too long, and getting longer, or I'd tell you a lot more wonderful things about the Braille Institute.

I have a closing story of ham radio's ability to create a new world, which I first read about in *WorldRadio,* a fine publication to which I am a contributing author. This story is of Ed Lewis, W5NOC, and his charming and dedicated wife Ellen, WA5KIY.

Ed was stricken with ALS (the disease which took the life of oldtime baseball star, Lou Gehrig). Over the years the ALS gradually disabled Ed to where his hands, arms, and legs became useless, and he could not speak because he couldn't move his tongue.

What we're talking about is an otherwise perfectly alert human being with emotions the rest of us feel, and the views we have, such as those which I am expressing at this very moment but am able to express only by virtue of God's gift to me. Ed, locked into an immobile state, could not even raise a view of the weather, of what he thought of the President's policies, whether the Dallas Cowboys might make it to the Super Bowl—all sorts of earthshaking matters which we all take so much for granted when we start a friendly discussion with someone. Ed had to keep such thoughts locked inside.

Until ham radio entered the scene.

Because Ed could still click his teeth together, they designed a clicker to slip between the teeth. With it, he was able to pipe code to Ellen just as neatly as could be. Finally, a voice to the outside world! Small matter, this? It all depends on which side of the fence you might be standing, does it not?

# · 36 ·

# Handi-Ham People

This is not a story of "handy" hams, though many HANDI-HAM members are quite handy. No, in this case, HANDI is short for handicapped.

The HANDI-HAM System is a special part of an organization in Minnesota called Courage Center. Courage Center works with thousands of children and adults who have physical handicaps such as vision, speech, or hearing problems. HANDI-HAM activities are one part of Courage Center, and a thrilling and unique side you should know about.

When I first wrote to HANDI-HAM Director Bruce Humphrys, K0HR, to inquire of the HANDI-HAM System, Bruce simply forwarded some pamphlets and a brief personal note.

That, of course, didn't satisfy me; I wanted more information. I wanted to know, for example, exactly what takes place when a handicapped individual reads about Courage Center and the HANDI-HAM System—someone who can't come to Courage Center but wants to get into ham radio. What does the HANDI-HAM System do? Here are the answers I received from Bruce:

> Understand, HANDI-HAM members are both the handicapped and the non-handicapped. They are either hams or they are the ones who want and need to be hams.

Here is what happens when a typical individual might read your book or hear about us one way or another. Let's say the prospective person's name is Bill.

Bill writes us a letter, usually stating his disability in general terms. We write back, explaining what the HANDI-HAM System does to prepare him for an amateur radio experience. We also send him an application form.

When Bill is registered in the system he is sent a welcome letter which details specific services available to him. We send out our basic amateur radio course, *Tune In The World With Ham Radio*. We may also enclose additional material, as well.

Immediately upon receipt of his application, we attempt to find a HANDI-HAM member or student member in his area who can work with Bill. If we have no one in the system living within a short distance from him, we attempt to find someone through other means such as the American Radio Relay League.

(By the way, we have a somewhat affectionate name we place on individuals who help others into ham radio: Elmer. It's a tag that keeps coming up, and, as usual, no one seems sure of how it got started.)

Phase 2 of the process as described by Bruce goes like this:

Bill can study at his own pace. All we ask is that he maintain some progress, that he doesn't give up. We try to contact our students regularly to see if they have any problems. At some point early in his studies, Bill will realize he can get excellent code practice by listening to actual on-the-air QSO's [ham conversations]. He will then send in a request to see if he can borrow an outfitted ham receiver.

Now we're getting specific. We've got Bill learning about ham radio. What next? Phase 3:

From this point until the time he is ready to take his exam, we are in a supporting role while the actual personal service is given by the trained radio amateur volunteer in his own home town. Once he passes the exam, he is eligible to receive a transceiver or a matching transmitter on loan from us. The equipment will be appropriately modi-

fied to Bill's handicap. We even send him 100 HANDI-
HAM QSL cards with his very own new ham call sign
on them!

In the final phase of the HANDI-HAM process, Bill gets
through the hurdle of his first on-the-air contact. He continues
to operate regularly until, soon, without even realizing as much,
he loses status as a student member and becomes a kingpin in
the HANDI-HAM System—Bill can now become an Elmer
himself if he wishes to. He will be ready to show another
handicapped member that it can be done. He can explain how
important it is to have those two qualities I always bring in—
interest and desire. Bill will be the first to point out, however,
that the path to ham radio for the handicapped is not an easy
one, that it involves hard work. But he will always say too that
once you have passed the exam, there's no other thrill quite
like operating your own station over the air.

So that's it, the mechanics of the fabulous HANDI-HAM
System. You handicapped readers are now at a crossroad of
whether to join us in this exciting and rewarding hobby, or not.
It's up to you. You are certainly entitled to your doubts, but
you will never know unless you take the first step of calling
Bruce at (612) 588–0811. Or write to him at: Courage HANDI-
HAM System, Courage Center, 3915 Golden Valley Road,
Golden Valley, Minnesota 55422.

By the way, the HANDI-HAM System has expanded its
activities to both coasts with camps, seminars, and other doings
perhaps closer to home for some. They'll tell you all about it.

And when you write, tell Bruce you are looking forward
to the day you QSO (contact) him on the ham bands. If you
live in my area of southern New Jersey, consider asking that I
be assigned as your Elmer. Of course—I couldn't research and
write about such a wonderful organization as HANDI-HAM
System without personally getting involved. I signed up as a
member/Elmer.

I like to talk about the handicapped in ham radio, but
converts can be pesky ramblers, and I don't want to carry on
too much. I am mindful of the preacher who compared a nice,
long sermon to a thriving lawn in which every blade of grass

had its place under the sun. The comparison dulled when someone in the congregation called out, "So, cut it short!"

And so I will, except to add a request to those who may
join the HANDI-HAM System from having first read this book,
that when you are licensed and on the air, I'd like to hear from
you. I'd like very much to be your first contact on the air
because I so much enjoy sharing the joys of ham radio with
new members, particularly the handicapped.

# ·37·

# YL Profiles

*QST* published an article on YLs (an abbreviation for "young ladies" but actually it means any woman in the hobby) from which I have taken these few, short profiles simply to give some views on what they think about the hobby and what they are doing.

Jan Shillington, N9YL:

> Amateur radio is so varied you don't have to be an engineer. I am not a technical person but I am an enthusiastic ham.
>
> I enjoy code and people. Ham radio is an outlet, an escape from the everyday world of children and muddy dogs. You turn the switch and you are no longer the harried housewife but a ham competing on an equal level of competition with every other person—man, woman or child.
>
> I know it's hard to learn and grow and be a housewife. That's why amateur radio is so great. You can watch the kids, stir the dinner and operate all at the same time. Using code is preferable, since any yelling or screaming at the kids can't be heard over the air.

Alice Moore, WB9LAD, expressed these thoughts about the amateur hobby:

There's nothing like its friendliness and friendship. There were evenings in high school when I would be studying at my desk and the temptation to turn on my radio would overcome me.

I knew that a whole world of communication was at my fingertips with the flick of a switch. Isn't that indeed a temptation? It's still just as exciting today.

To me, friendship is the big thing in amateur radio. It's a way to have more sunshine in your life and a very good way to pass some on, too!

Dr. Chris Haycock, WB2YBA, admits that at an early age she was more comfortable on a baseball diamond than on a dance floor: "If a young girl is physically large or awkward on the dance floor, over the air she's the same as anyone else. Amateur radio offers her a chance to compensate. If she's shy or introverted, she can make hundreds of new friends, easily."

Sue Smith, K3YL, says she is the only ham in her family: "I didn't have any outside influence. When I saw what amateur radio was all about, it was love at first sight. Here was a hobby in which I could use electronics and talk around the world at the same time. The combination is fantastic."

Flo Majerus, W7QYA, is a pilot, teacher, and world traveler. "The contacts amateur radio gives you are great. You have a friend every place you go. I got off a plane in Russia, once, and someone was right there waving my QSL!"

One of my local YL favorites is Jan Scheuerman, WB2JCE. Jan served as president of the YLRL—Young Ladies Relay League—in 1971.

Jan was living in New York when she successfully petitioned the governor to proclaim a "YLRL Week" in New York State. Arthur Godfrey, who happened to be radio amateur K4LIB, gave Jan a big part on one of his radio shows in which they talked quite a bit about YLRL week and YLs in ham radio.

One of Jan's first contacts on the air was with radio amateur WB2IGD, Hugo, who lived near Atlantic City. They had QSOs on the air daily for years. They finally met and got married.

# · 38 ·

# B as in Bravo

This unique but true story of amateur radio was given to me by a very unusual amateur radio operator. The story to follow is in the first person, somewhat as told to me by radio amateur WA2IPM, Brother Ben, as he is known on the air.

Brother Ben shortened his name from Bernard to Ben in a joking moment when he found it too bothersome on the air to spell out the name Bernard with the phonetics we use—"B as in Bravo, E as in Echo, R as in Romeo. . . ." Here's Brother Ben's story:

> I am Brother Bernard (Ben) Frey, a Capuchin Francis-can friar. I am also radio amateur WA2IPM, a fascinating addition to my life for which the friary has never been quite the same, as I will attempt to describe.
>
> There was the time, for example, when I was erecting a small ten-foot tower, which doesn't sound as though it should really have been a chore even though I was working alone. I, however, was installing this tower along the edge of a four-story penthouse roof. Things did not go quite as I had planned.
>
> To make what I consider to be a long and painful story short, I was working with the guy wires and happened to let

go of a tangled wire at precisely the wrong moment. In a flash, the guy wire, the tower and I went hurtling over the edge of the roof!

I don't know why I wasn't killed right then. Probably because it wasn't yet my time. When I regained consciousness, I found I was in a most awkward position. I had not fallen to the ground but was hanging upside down in mid-air with my toes wedged firmly between the chimney eaves!

I must have hung that way for about an hour before someone finally heard my feeble calls for help. Shortly afterward, our fire department came to rescue me. They had to saw off the eaves because my toes were wedged in so tightly. They then lowered me to the ground—bruised, bleeding and embarrassed but thankful to be alive.

In 1970, my Order sent me to Honduras to install five amateur radio stations for the International Mission Radio Association (IMRA). IMRA is a group of radio amateurs in dozens of countries throughout the U.S., Central America and South America; many of them are clergy. They serve to provide services to missionaries, volunteer workers and people of all denominations. Their motto is "People helping people."

At our first stop in the tour, at the town of Ocotepeque, our installation of the first station went quite well. Power was readily available and the station was installed in a jiffy.

San Marcos, the next stop, proved to be a bit more interesting when I found that the beam antenna had to be carried up one of those dangerously sloped Spanish-tiled roofs. Would you like to try that stunt with an awkward beam slung over your shoulders? I had my moments, that day.

As in most Honduran towns, San Marcos provided electricity for only about two nighttime hours. When it came on the first night, the electrical voltage level swung wildly from 50 Volts to 200 Volts (and 115 Volts is normal). Fortunately, I had brought along regulators which were used to hold the electrical level steady.

What I had not brought along, however, were television interference (TVI) filters. Much to my surprise, I overheard one of the parishioners complain of some mysterious interference to their television reception. How was I to prepare for that? I didn't even know they had television down there.

By the way, I was accompanied on this tour of Honduras

by Dr. John Schmidt, who later became radio amateur WB0ACU. Dr. John was along for the ham experience, although he didn't really expect one such as he encountered at our next site of Santa Rosa de Copan.

I remember that evening in Santa Rosa de Copan quite well. We were all busily engaged in the checkout of the beam antenna—Dr. John, the Bishop and myself—when suddenly there was a tremendous thunderbolt. A lightning strike instantly hopped from the control box, right through my hand. Dr. John claimed that it struck him directly on top of the head. The Bishop simply turned white as a sheet. Yes, we do remember Santa Rosa de Copan. How could we ever forget?

At the next village, Guarita, we found our only convenient spot for assembling the beam to be in the town square. Our activities resulted in quite an audience, including the ever present soldiers. We worked diligently, I assure you.

Generators for Guarita's electricity were conveniently located at the church. Such a handy location gave us direct control of our own and, of course, the town's electricity.

In order to check out the equipment, we had to turn the electricity on during daylight. We soon found out the consequence of doing so. You see, the entire village turned out, thinking it was a fiesta. Electricity during daylight? It had to be in honor of something or other, and no one wished to be left out.

Until that point in our tour we had always enjoyed the convenience of bus or truck for our transportation. But the best was saved for last, and La Incarnation was the very end.

It was a 9-hour trip by muletrain over mountain trails which no mechanical locomotion could possibly survive. Once was enough; our prayers were answered for the return trip when an American helicopter took us back in a passage that was a delightful experience, looking down and sneering at those mountain trails which had been such trouble crossing on the way in.

What I remember most about La Incarnation (besides those blisters on my seat, thanks to the muletrain) were those who watched us with wondering eyes as we assembled our equipment. In particular, there were the little children asking such questions as, "Is it true you Americans live underground?"

So it was that we completed Operation Honduras, and none too soon, as it turned out. After returning to America I was in contact with Tom, W9LII, who was in Honduras at the time just as Hurricane Fifi struck. Tom was reporting the violent winds and a fantastic plunge in barometric pressure . . . and that was it, as the hurricane knocked out all power in Honduras.

I remained at my ham station in the States all that week, directing and relaying messages and phoning the needs of those poor people to the authorities. Those ham radio stations we had installed were the only means by which the Honduran government was made aware of the extent of damage and the needs of the people. It was all a very trying time.

There was much more to Brother Ben's story of the hurricane—the lives saved, the sick aided, the lonely cheered, and the inevitable death messages delivered. His charities of time and personal sacrifice through amateur radio were supported by the other generous radio amateurs who were and are involved in similar activities. Brother Ben best expressed matters when he closed with, *"The rewards of amateur radio have blessed my life many times over."*

# · 39 ·

# Mr. DX, Ross Hansch, W9BG

Ross Hansch was born a DXer (foreign country contacts). It took him all of his first dozen years on earth to realize as much.

In 1921, on his way home from grade school one day, he stopped in at the electrical shop his Dad owned, and that was when he heard radio for the first time. That was also when an extraordinary DXer was born because one listen to stations coming in from "hundreds and hundreds" of miles away was so fascinating to him that it set a determination he didn't realize at the time—to become the world's foremost DXer. And that determination came to him with as subtle a hint as a tornado-driven projectile takes its mark. He was hooked for life at that instant. It happens that way to some of the true DXing radio hams.

Sixty years later, Hansch moved about the radio airways throne of DXing with a sureness few others could match. For some time, he had shared the DX spotlight with another expert; both of whom had logged contacts with 365 stations for the ham radio all-time world record. Funny thing, but they both had missed an obscure country that permitted amateur radio

for only one day, and both were shy of the perfect mark of 366 for a long time.

When I had asked Ross for a peek into his lifestyle as a dedicated and celebrated DXer, he replied,

> There's so much going on when you start talking about what's happening in DX and all the people involved, it's just a hopeless thing to get into in a short discussion.
>
> For example, take the DX program at our ham club, which we aim toward beginners. You really can't relate to them in one session. You've got to talk the language and the language comes only from experience. That's the key point.

Did he spend much time at DXing?

> The problem with DX as a hobby is you get so engrossed you start associating with a lot of people, local and otherwise, in DX, and it gets to be an obsession. It starts to eat at you like a cancer. Frankly, the advice I give to anyone who comes to me and says they'd like to be a *real* DX'er is: Don't.

Yet it was obvious in talking to Ross that he never would have spent his lifetime of hamming in any other way.

We talked on the air one day. Interestingly, the band was a bit crowded at the time, and Ross excused himself for a minute; I listened off frequency to hear him call into the conversation taking place to explain he had a contact nearby and would appreciate if they would hold off for a whole. Everyone knew him, of course, and courteously honored his request. (No ham has any exclusive right to any frequency.) I was impressed with their obvious respect for this superb ham radio operator and DXer.

Ross spoke to me rather enthusiastically about his first-name basis with radio hams just about everywhere in the world—thousands of them. And he talked about them the same way we might talk about the gang at the office. I soon got the feeling that his half-century of activity in the international sport of ham radio DXing permitted a comparison with the way our Secretary of State operates. While they may talk to the "big boys" in a dozen or so countries, Ross has talked to thousands of ordinary but just as important people in hundreds of countries. I'd vote for Ross any time.

W9BG also proved that you don't need either the total bankroll or tons of complicated equipment in order to work and enjoy DX. In fact, he started out by building his own gear, back in 1920, and he kept on building his own until about 1958 or so, when he finally went partly commercial with some of his equipment by buying a radio kit. "I've always been a tinkerer," Ross told me. He went on to say:

> I used to have a 50-foot windmill tower which made it convenient to experiment on antennas. In 1954–55, at the bottom of the sun-spot cycle and when DX was poor, I had eleven different 20-meter beams up in the air.
>
> Then, when trapped antennas [see glossary] came into use, the whole world of antenna design changed. I did an awful lot of experimenting with them.
>
> DX'ing surely has been a love of mine all the time, but it has cut into my fun at experimenting, at rag chewing and at just about everything else under the sun.

You recall my earlier guess of how real DXers either never sleep or they get jobs by which they can work on the radio during the wee hours of the night? Ross fit into the pattern: "I worked in a broadcast station where, naturally, different shifts opened up for us that gave many opportunities to be on the air at various odd times. Of course it's true that many of the real DX'ers have jobs like that."

Pure DX—the art or science of turning all your energies into contacting more and more foreign countries—is a strange and fascinating world. It can't be described in a few pages; Ross had said he could fill a 500-page book on the subject while still barely scratching only a few of the DX surfaces. "But no one would read it. Reading about DX is not what the DX'er wants or needs. *DX has got to be lived.*"

And that's just about the size of it, coming from someone who really knew the inside story of DX as spread out over more than a half-century of operating. In his last letter to me, Ross wrote: "I sort of lived in the dark ages, using old equipment. The reason for my high DX score is because I've been around almost forever, have a lot of helpful friends, and have worked just about everything that came along."

Plus, of course, having had loads of interest and desire.

# · 40 ·

# World's Youngest Ham

If you recall, I told you earlier about a record for the lowest age in getting an amateur radio license. The hero of this chapter is Guy Mitchell, WD0DVX, who passed the novice exam at the ripe old age of 4. That's right, a 4-year-old passed the amateur radio test. How about that?

I considered it such a neat trick that this little expert had passed the exam that I looked into it to see if he was pure genius or was anything like the rest of us. Turns out he was mostly like the rest of us but, of course, somewhat better because at that age you have to be pretty good to take any kind of written exam.

I wrote to Guy's parents for his ham radio profile. Both Guy (who was then 7) and his mother, Cindi, who is radio amateur WD0DFU, replied. Guy's letter was a treat! The envelope proudly carried the title, "World's Youngest Ham."

Carefully written out, Guy's words to me, exactly as he wrote them, were:

192

I got interested in ham radio when my Mother took a ham class in Feb. I listened to her practice. I looked up letters I didn't know. Mom, Dad and I would send code. It was easy for me to learn.

The theory was hard for me to learn. Since I was reading since I was three, my parents thought I was ready to take the test since I had such an interest in the code and enjoyed doing things my parents did. By the end of Apr. I took the written test, just before I was five. Now I am seven, soon to be eight.

*Kids should keep trying.* It's fun in the end. They should ask any questions they have. Do the code with a friend or parent.

There you have it. Straight from an "old-timer" who, at age 7, had already been at the hobby for 2 years.

Cindi's letter was equally interesting. She described Guy's interest and progress this way:

The hardest part of code for Guy was to be able to copy fast. He was just learning to print and his fine sensory motor skills were not fully developed. His father sat nearby giving assistance with writing.

Many of the people Guy has contacted on the air could not believe he was five, since he was very good at sending code. He enjoyed it a lot and practiced often. It was such a joy watching him learn the code so quickly. We sent code to each other any free time we had.

Guy would tire of QSO's that lasted more than 30 minutes. QSO's that lasted only 15 minutes were more timely.

He liked doing things we were doing but after he started school his interests changed to what his peers were doing, such as football.

We enjoy taking part in hamfests and belonging to the local ham club.

Cindi also talked about how the radio theory was more difficult for Guy than most because his vocabulary was too limited. For example, when his Dad would try to explain to Guy what electric current flow was, he would first have to teach him what the word *flow* meant.

In the process of learning about ham radio, Guy was also taught how to use a calculator to do multiplication and division. I have mentioned in earlier chapters that there is very little arithmetic in the novice test, but for this little novice the idea of adding numbers was something he had to learn from scratch.

"Guy had a lot of patience," Cindi wrote, "and a long concentration span, unusual for a child his age. He had the determination to get his license, and he worked at it."

"Determination" and "worked at it" are, any way you slice it, another way of saying what I've been saying all along: interest and desire. That's what it takes to get into ham radio, and with those qualities, even a 4-year-old can do it.

# · 41 ·

# Contest-Covering Capers

One of the other books I've written is a book called *Guide to Unusual Contests in America*. This book is not about the kinds of contests for which you send in box tops, not at all. It is a collection of descriptions of America's whackiest contests: The national hollerin' contest, the national cow-chip-tossing contest, tobacco-spitting contest, chicken-flying contest, chicken-plucking contest . . . you get the idea. You usually see several of these on the TV news and on various TV programs when they come up.

One of the contests in the book—the very last one, in fact—is on amateur radio's Field Day contest which I considered to be unusual enough to qualify for the book (and I think you will agree when you get a radio amateur license, meet the local hams, and get a bit involved with what they do). Every ham club in America, it seems, takes to the woods on Field Day, and that's what it is—a weekend of operating in the field under emergency power such as batteries or gas generators, with equipment we own, trying to make as many contacts with other ham radio stations who are also out in the woods that weekend. I've mentioned Field Day before, and it bears repeating because it is in keeping with one of the oldest amateur radio traditions and one of the most enjoyable activities we have.

Getting back to my book and to those whacky contests, one that comes up the weekend before Field Day, in early June, was started by a radio amateur—this is the National Hollerin' Contest, to which as many as 10,000 spectators show up.

The National Hollerin' Contest is held at a country crossroad named Spivey's Corner, in North Carolina. And a crossroad it is (we zapped by it twice while looking for it the year we went there).

Radio hams are everywhere, as you get to learn when you get into the hobby. At the Hollerin' Contest, they had a "ham holler-in" in which they had set up a complete ham station on the contest grounds. If you contact their station that day, they send you a very attractive QSL card. My son Jim was with us when we attended, and he took a turn at operating their rig. Surprisingly, his first QSO was with a local YL ham from back home in South Jersey. We all got a kick out of that. But, as I keep saying, ham radio has more kicks to it than just talking on the air to strangers—a lot more.

Another unusual contest is the Ramblin' Raft Race, sponsored by the American Rafting Association. It is held in several cities throughout the United States, the first one of which was on the Chattahoochee River in Georgia in 1969. And from that first race it has grown to where these races now bring in over 400,000 spectators and 50,000 participants. The *Guinness Book of Records* lists this as the sporting event with the most participants in the history of the world. That's impressive.

So how does amateur radio feature in this? Quietly, but very effectively, I assure you.

The problems of handling 400,000 spectators get to be staggering, as you can imagine. They are made easier by the contributions of several local radio amateurs who set up a communications network for the action.

When I contacted the radio ham who had been a coordinator for the affair, he sent a copy of a 23-page document he had written called, *Operational Plan, Amateur Radio Safety Network in Support of the River Raft Race.* Anything with a name like that has got to have something of interest in it for everyone, so let me tell you about it just as an idea of how we hams get into just about everything; and as another idea of how we do more than just talk on the air.

For example, there were sections in the plan on assignment of operators; briefings; contacts; credit and publicity; control; traffic procedures (traffic for messages, not for cars); debriefing and comments; disclaimer; hassels and QRM; identification; materials distribution; operating frequencies; parking; personal comfort; schedule; and station locations. Thought of just about everything, didn't they? I liked the section on comfort, which said: "The rafters will be out in huge numbers, rain or shine, and the safety of amateur radio network will be vital 'weather or not.' Please plan accordingly to protect yourself and your equipment. Protective hats, clothing, umbrellas are recommended as are folding chairs and card tables. Food, drink, and toilet facilities will not be available at most locations where amateurs are stationed. A well-stocked picnic cooler will handle food and drink problems. Good luck on the other."

Rob Diefenbach, WD4NEK, who had a big hand in a lot of the preceding "operational plan," is an amateur with the hobby sideline of amateur radio support to local community events. He has also handled the Atlanta July Fourth parade and the famous Peachtree Road Race where 25,000 runners compete in a mostly uphill 6-mi road race.

As Rob told me,

> Unless these things are done with the greatest attention to detail, they just aren't worth doing. Amateur radio rides with these events and we like to do our very best. All we get out of it is a lot of hard work, the satisfaction of working together, fun, good times, memories and last but not least, great publicity for the hobby!

I really couldn't have said it any better.

If you happen to be near where one of these raft races are held (Atlanta, Jacksonville, Washington, D.C., Dallas, Houston, Louisville, and a few other places), consider taking a look at them. And when you have your ham radio ticket, consider joining a local group that provides free communications via ham radio for such an event in your own area. Believe me, it's a fun activity, it's a great way to widen your circle of friends, and it's a memory to treasure.

By the way, I took in the Ramblin' Raft Race one year when it was first brought to Philadelphia. I can tell you I was really impressed with what I saw. I covered it for my contests book, and was treated, along with the family, to a view of it all from a yacht which cruised up and down the river while all sorts of wild structures (some not really too seaworthy) went floating by. Since it was a hot and humid day, those rafters who were half in the water all the way downriver were among the winners even if they didn't make it to the finish line. At least they kept cool!

# · 42 ·

# For Chess Players

"From little acorns do mighty chess trees grow," is a distorted version of a saying that may seem to have nothing to do with amateur radio except for the fact that the acorn is a nut and, some say, so are we chess players, particularly those of us who meet to combine hobbies by playing chess over the air.

The acorn in this story is an international organization of chess players who are now playing games on the amateur radio bands in many parts of the world.

Chess & Amateur Radio International (CARI) is the acorn that was planted because of my son Jim's amateur radio interests when they expanded into chess. Jim tried hard to find other chess players on the air who were radio amateurs, but he had little luck.

I wrote a letter to *Chess Life* magazine, the official publication of the U.S. Chess Federation. In that letter I talked about the joys of combining chess with amateur radio. I expected, and received, letters from their readers who were radio amateurs. (I knew they would have several chess-playing radio amateurs in their membership.) But I wasn't prepared for the many nonamateurs who were chess players but also wanted to become radio amateurs.

Hundreds of letters poured in. The one that fired the shot heard around the world was the one that asked, "Why don't you organize us?" We did. Jim and I started Chess & Amateur Radio International, CARI.

CARI members now regularly meet on the air to talk about and to play chess. I'm no expert at chess; far from it. But I can tell you it's fun to work and play with such dedicated people. Radio amateurs are dedicated people, as are chess players. Put them together and you have a marvelous group of super people. Anyway, when we get together on the air, and after all the talking is through, that's when we get into the games of which there may be a half dozen going on at the same time. It may all sound strange to non-chess players who listen in, but, to chess players, it's pure heaven.

We started CARI in 1982 and, in our first year we had nearly 150 members, some from as far away as Australia and New Zealand. We had tons of publicity, not only in the amateur radio publications but also via the Associated Press when we made the "big time" via a radio chess match held at my house with six schoolkids playing one of our CARI members in Massachusetts, John Dould, N1BHL, who is also CARI member Number 3 and CARI's tournament director. John, being so much better than the kids who had learned chess only a few months earlier, spotted each of them his most powerful chess piece, his Queen, as he played them all—six games at the same time.

John's famous quote was carried by AP throughout the country when he announced over the air, "I must have been nuts to agree to this!"

Why? Well, in two hours of playing, he had lost two games, had draws in three, and one was incomplete because our player had to leave for baseball practice. It was a fun time enjoyed by all.

These days, we play games with radio amateurs in Europe, Australia, and New Zealand, with members who all agree that CARI is the best thing to come along since the wheel. Every week newcomers to CARI add their blessings that there is finally an organization such as CARI that brings these two fine hobbies together.

We also have non-radio amateur members in CARI, some of whom are now licensed radio amateurs. They "saw the light" and hurried to get their licenses. We helped them along. How? With the best book we know by which to introduce people to the hobby of amateur radio without scaring them away—*Amateur Radio, Super Hobby!* They had the interest and desire; we showed them the way.

Let me tell you that chess by amateur radio is entirely different from any other brand of chess, whether you're talking about sitting down opposite an opponent, playing chess by mail, or playing against one of the computers. By radio, the other player isn't in the same room and may not even be in the same country.

In fact, I've played chess with my very good friend Kirk McMillan, ZL4PX, all the way around the world in New Zealand. I've never beaten him at a game (I never said I was an expert at chess), but we certainly have had fun at it. My excuse for losing to him has always been the hour—skip conditions are best in New Zealand on 20 m at the unhealthy Eastern U.S. time of 2 a.m.! Our Australian CARI members check in also—Kevin Moore, VK3ASM, and Craig McMillan (no relation to Kirk), VK3CRA.

(By the way, I've used that term *skip conditions* several times in this book. I hope you understand its meaning because one of the questions on the novice exam is, "What does the term *skip* mean?" And another is, "What is the meaning of the term *skip propagation?*" You are likely to get one of these in your exam.)

Back to chess: Picture yourself playing a game of chess halfway around the world with someone you have never met in person, someone whose chess pieces move across your board almost as if by remote control, someone who breaks down your best strategy as they cunningly scheme against you, again and again. I tell you, if you like chess, you will absolutely love chess by amateur radio. (The address for CARI is in App. 1.)

# · 43 ·

# George Grammer, W1DF

Sixty-five years in amateur radio is our story on one of the hobby's grand old men, George Grammer, W1DF.

Of those 65 years, George spent 41 at the American Radio Relay League headquarters in several different positions, mainly as technical editor of the ARRL publication, *QST*. Here is the ham radio profile of a member who has truly dedicated his entire life to the hobby.

At the age of 8, in 1913, the youngster from Philadelphia had a passionate yearning to learn the Morse code and a fascination with railroading. These interests led him to acquire a Morse practice set by which to learn the code, but his progress was slow because there were no others around with whom to practice the code.

In 1915 he came into possession of a free catalog which listed the wireless equipment of the time. "That 1915 catalog was a treasured bible to me for quite a while," George wrote to me, "even though I was but ten at the time and had absolutely no money to spend."

Christmas of 1916 brought a small cash windfall to him,

money he promptly used to begin his career in amateur radio and experimenting with radio equipment. He purchased parts for a crystal receiver.

Unfortunately, World War I was underway before the parts arrived, and during the war years the government did not allow people to have either transmitting or receiving equipment. "My father had to get permission from the Commandant of the Philadelphia Navy Yard to have them delivered at all," George recollected.

With restrictions removed after the war, George put together the crystal receiver. Then, with the station at the Navy Yard transmitting code practice for amateurs, he managed to build up his code speed. And after acquiring a Ford spark coil and gap (see glossary), he went on the air with his first station, such as it was. "Even though I had no license at the time, a spark coil wasn't supposed to have enough range to reach from West Philly into New Jersey. And we had been told the regulations said a license was not required to work within your own State."

Fate again stepped into his life via a visiting cousin who was a radio amateur. The cousin also happened to have a copy of *QST* with him, and that did it for George. *QST* seemed to cover just about everything he was interested in and wanted—descriptions of radio activities, who was doing what, and even lists of the call signs of other radio amateurs.

From that early assist, plus his working with a few other boys his age who were also interested in ham radio, George built up his code speed to the then-required exam speed of 10 wpm.

I recall my immense personal satisfactions during my beginning period in amateur wireless. There was, for example, the first time I ever heard a signal on the air coming through my own home-built crystal receiver. There was also my very first contact using that old Ford spark coil. And there was the thrill of my very own license.

Also, of course, were the pleasures of hearing DX such as 3EN in Norfolk, 8XE at State College, Pa., 1HAA and 1AW in New England and 9ZJ in the midwest. All this on my little old crystal detector!

Later DX improved considerably for George when he built a tube-type receiver using the new RCA Audiotron vacuum tube. "It opened an entirely new world of receiving to me. And then when I also found out how to make it regenerate, it was like stepping onto the Promised Land!"

After using the old Ford coil as a makeshift transformer (see glossary) for his first tube-type transmitter, a subsequent rig with a real plate transformer truly opened DX contacts for him beginning with his first European contact, F8GO, in France, in 1925.

George joined the *QST* staff as a technical information expert in 1929 and became assistant technical editor the following year.

> Part of my job was to work on technical problems of interest at the time, as well as to develop equipment of appeal to *QST* readers, particularly newcomers. The gear had to be inexpensive in those days as we were in the middle of the depression. This job was exactly the sort of thing that appealed to me.

Take, for example, his 10-m pioneering work. In 1929, no one knew much about the effects of sun-spot activity on radio communications. The new 10-m band seemed to be little more than a barren wasteland. Not to be discouraged, George's old call sign, 1DF, was regularly heard calling "CQ ten meters" in a personal crusade to stir interest in this band.

Not until about 1935 did the 10-m band come alive, and, when it did, it brought in DX in a way no ham had ever before experienced. Even with low power, the amazing 10-m skip conditions brought in South Americans, South Africans, Australians, and even Asian stations. Yes, the vagabond 10-m band had finally roared back to life as the now-understood mysteries of the 11-year sun-spot cycle began to peak for the better.

After World War II, George returned to *QST* and into a new world of amateur radio operating. Now, the "one-eyed monster" was popular—television and, of course, television interference. (We call it TVI from which we obtained the handle, "one-eyed monster"—the "I" in TVI—see glossary.)

Although there had been instances of interference from

amateur radio transmitters before the war, such radio interference was more easily solved. But TVI was different; now it
was a case of what the eye saw on a TV screen that was being
affected. TV receivers of those days were so poorly designed
they picked up interference from just about anything that made
a noise. TVI was, indeed, a new challenge. And the solutions
to those challenges were to become the personal crusade of
W1DF. He did much to advance the amateur art in this matter.

Regarding his years as a radio amateur and technical editor
of QST, George added,

> My satisfactions were derived from the publication of
> articles that, I hoped, would help the average hobbyist to
> keep up with things.
> To me, this has always been the really rewarding part
> of the work.
> In my own ham operating, getting results by using simple
> gear and simple antennas have always been my first objec-
> tives. For practically all my ham life I have used transmitting
> power under 100 Watts. I never did have a beam antenna,
> only wire antennas.

Retired in 1970, W1DF is seldom heard on the air these
days due to health problems. "I still listen in, though, and I'm
still as fascinated with code after all these years as I ever was."

This same fascination, beginning at age 8, was what George

George and the one-eyed monster.

offered amateur radio; with which he helped bring the hobby through its infancy and into the era we now face in so many amazing technical spin-offs—transistorized equipment as small as a cigar box that once was built into a relay rack 6 ft high; computerized transceivers that remember who was called; automatic code sending and receiving sets; amateur television; satellite communicators. The list grows every year, but they all had their beginnings in the 1DF era, the era and society of the Iron Men to which George Grammer paid his dues.

# · 44 ·

# We, the People . . .

*QST,* the publication of the American Radio Relay League, lists a feature called *Strays* in which are found various tidbits under the heading, "I would like to get in touch with (radio amateurs). . . ." I found these items to be a fascinating collection of who radio amateurs really are, expressed in a way no other definitions could.

Here, then, is a collection of several *Strays* from *QST*s. I've saved this chapter for now simply because I consider it a fitting closing, this different description of what I've said all along. We are everyone.

"I would like to get in touch with (radio amateurs) . . ."

". . . who are hams, in a community theater, that is."
". . . especially women, to join a stitch-and-chat net."
". . . with family members who have pacemakers."
". . . who are in real-estate organizations."
". . . who are members of Charismatic Christian communities.
". . . who are born-again Christians.
". . . who served in the 136th Signal Radio Intelligence, World War II."

". . . who are members of the Photographic Society of America."

". . . who are involved in the Right-to-Life movement."

". . . teenagers to play chess, checkers, battleships, etc., on the air."

". . . industrial electricians."

". . . air traffic controllers."

". . . who are rose growers."

". . . aviation veterans of World War I."

". . . broadcast band DX'ers."

". . . who use amateur radio for model rocketry."

". . . who have had open-heart surgery."

". . . willing to send electronics books to a prison inmate."

". . . who are amateur astronomers."

". . . who are pediatricians."

". . . members of SPEBQSA—Society for Preservation and Encouragement of Barbershop Quartet Singing in America."

". . . willing to rent house and ham shack to visiting ham family from Africa."

". . . to start net for hunters, shooters and fishermen."

". . . former weather bureau employees."

". . . who participate in drag racing or custom car showing."

". . . who are square-dancing fans."

". . . who hold patents."

". . . 13 years or younger."

". . . interested in space colonization."

". . . who are kayakers or canoeists."

". . . to establish a Free Mason amateur international group."

". . . to start a Disabled American Veterans net."

". . . who show purebred dogs, especially terriers."

". . . who collect metal robot toys."

". . . involved in any way with the art of prestidigitation."

". . . who were in the radio gang aboard the USS *South Dakota*."

". . . who are stamp collectors."

". . . members of the Good Sam RV International Club."

". . . from India, living in the United States."

". . . newsletter editors, to meet other lonely editors."

". . . to form an International Rotary Club net."

". . . with experience on aircraft ham antennas."

". . . to form a Veterans of Foreign Wars net."

". . . interested in joining Amateur Scientists Radio Organization."

". . . interested in long-distance running."

". . . who are police officers, sheriffs, judges, or gangster buffs."

". . . to establish an international net for seafarers to talk to families."

". . . who work in naval architecture."

". . . who work with rural or agricultural development projects."

". . . interested in unusual structures such as geodesic domes or yurts."

". . . Orthodox Jews."

". . . in commercial broadcasting such as DJs, program directors, etc."

". . . who are also twins."

". . . Pan American Airways employees."

". . . interested in historical photographic processes of carbo, oil, bromil, etc."

". . . who have devised ham antennas for mobile homes."

". . . involved in solar electric energy conversion."

". . . interested in long-distance balloon flights."

". . . interested in forming a solar flare network."

". . . who have used balloons to raise antenna arrays."

". . . to bring a dead language back to life talking code in Latin."

". . . who are physicians or teach biomedical instrumentation."

". . . who speak French or Spanish, to help students develop skills over the air."

". . . willing to exchange cars and home for one year with Australian ham."

". . . active in diplomatic or consular services."

". . . who are yachting enthusiasts or are in amateur radio clubs which operate from yacht clubs."

". . . who were in the original Royal Order of Hootowls."

". . . Franco-American hams in New England for Sunday afternoon ragchews in French."

". . . who speak Italian and are interested in forming a net."

". . . from other Explorer posts majoring in amateur radio."

". . . who heard Amelia Earhart during her last flight in 1937."

". . . who are antique wireless collectors."

". . . interested in hydroelectric power."

". . . Armenian-American Hams for ragchew in Armenian."

". . . who are breeders or exhibitors of pure-bred cats."

". . . *who became radio amateurs after reading this book!*"

# · 45 ·

# Test Time!

Here is a little quiz for you, just to prove to yourself that you actually have learned quite a bit about amateur radio from this easy-reading book that really wasn't meant to teach.

Keep in mind that these questions are not necessarily on the FCC novice test.

Check yourself out. If you answer about half of these questions correctly, you know you can soon become a licensed radio amateur. We already know you have the interest, or you would not have read this far. So add a touch of desire to it all, and then I'll know I will hear you on the ham bands one of these days.

Answers to the questions follow the quiz.

### *True/False*

1. A license is required for every level of amateur radio including the novice class.

2. The only requirement for a novice class license is to be a citizen of the United States.

3. The minimum age for being allowed to become a radio amateur is 18 years.

4. All amateur radio exams are given at FCC field offices.

5. Only electronics engineers or technicians can hold amateur radio licenses.

6. The Federal Communications Commission was established in 1896, when radio communications was first started.

7. One function of the FCC is to regulate amateur radio activities.

8. The American Radio Relay League issues amateur licenses.

9. Television transmissions are permitted on the amateur bands.

10. Radio amateurs can talk via satellite.

11. Russian amateurs use special codes to keep amateurs in other countries from using their amateur radio satellites.

12. The best amateur band for working foreign countries is the 160-m band.

13. The FM broadcast band is higher in frequency than the amateur HF bands.

14. If one kilohertz is 1000 cycles, then one megahertz is 1,000,000 cycles.

15. The widest HF amateur band we have is the 10-m band.

16. The 10-m band is most active between midnight and dawn.

17. All licensed radio amateurs are required to take part in relaying messages.

18. A *traffic net* is a scheduled meeting of radio amateurs for the purpose of relaying messages.

19. Members of traffic nets must sign off the air at the end of their net.

20. Playing chess over the air is not permitted.

21. WAS means "Write All Stations."

22. The WAS certificate is obtained by applying to the FCC.

23. QSL means to acknowledge receipt.

24. QSL cards must have your photo on them.

25. It is not legal to send QSL cards to a foreign country.

26. Taking part in amateur radio contests is optional.

27. Call signs are not used by novices.

28. All American call signs begin with the letters W or K.

29. The continental United States is divided into 10 call sign districts.

30. When radio amateurs move to another state, they must apply for new call signs.

31. DX is an abbreviation of the word *distance.*

32. The DXCC award is received for having confirmed contacts with 100 different countries.

33. DX conditions are generally best between midnight and dawn.

34. A DX-pedition contact does not count as an official DX country contact for a new country.

35. The leader in DX contacts has contacted only as many countries as are in existence today.

36. *Propagation* is a technical term which refers to the path of radio waves in the atmosphere and above.

37. Radio signals in the HF band are reflected back to earth quite readily.

38. The ionosphere remains at the same distance above the earth day and night.

39. Sun-spot activity has no effect on the propagation of HF signals.

40. All amateurs are required to participate in emergency preparedness by having emergency-powered equipment.

41. The purpose of the Field Day contest is to make the most possible contacts in the shortest time.

42. Field Day stations can have as few as one member.

43. Public service is the first basis of amateur radio.

44. Good eyesight is necessary in order to become licensed as a radio amateur.

45. Members of congress are not permitted to have an FCC amateur radio license.

46. Operating DX is a good way to learn about geography and other countries.

47. Amateur stations are not permitted at sea.

48. Marconi became famous by being the first to hear a wireless signal from across the Atlantic Ocean.

49. Women are not permitted in amateur radio.

50. YL means "young lady."

51. OM means "open message."

52. YLRL is an organization of U.S. amateurs only.

53. XYL refers to a wife.

54. Only one member of any family is permitted to hold an amateur radio license.

55. Repeaters are permitted for use only to relay messages.

56. High locations are best for repeaters.

57. Most repeaters are restricted to operation by local members.

58. Walkie-talkie units are too low in power to be useful with repeaters.

59. It is not possible to contact a station more than 500 mi away with a walkie-talkie.

60. You should not use a repeater to call for a hotel reservation.

61. A code test is required for every amateur radio license.

62. The maximum code speed permitted on the air is 20 wpm.

63. Being "code happy" means to successfully pass an FCC code test.

64. Code used by radio amateurs is the International Morse Code.

65. Code speed requirement for the novice class license is 5 wpm.

66. Code ability is not required for the technician class license.

67. All amateur radio licenses must be renewed.

68. Amateur radio exam questions are usually of the multiple-choice type.

69. Novices are not permitted to contact foreign countries.

70. The top grade amateur radio license is the advanced class license.

71. Applicants for the novice exam are assigned call signs when they take the exam.

72. Most questions on the novice exam deal with rules, regulations, and operating procedures.

73. Novice exams require both code and theory testing.

74. The flow of electricity is called voltage.

75. Resistance is measured in units of ohms.

76. A 10-V battery connected to a lamp which draws a current of 5 A has a resistance of 5 Ω.

77. A glass soda-pop bottle is an electrical insulator.

78. Novice exams are graded at local FCC field offices.

79. RST 595 is a very good signal report to receive.

80. Amateur equipment made before 1950 is not legal for use on the amateur bands anymore.

81. A *transceiver* is a transmitter that does not have a receiver built in.

82. Novices can operate by voice only in the 2-m band.

83. Beam antennas cost less than wire antennas.

84. Your house must be rewired for power when you use a ham radio transmitter.

85. Licensed novice amateur radio operators are permitted to operate transmitters they have built themselves.

86. A home-brew rig is an illegal transceiver.

87. Novice class operation is permitted on 7.123 MHz.

88. The *ARRL Handbook* is strictly a guide for FCC rules and regulations.

89. Soldering irons are used to connect electrical parts in a circuit.

90. *Direct current,* dc, is current you get directly from the electric company.

91. A worldwide DX contest is sponsored by *CQ Magazine*.

92. Novice courses are conducted by the FCC.

93. A short circuit is a circuit in which all the parts are close together.

94. Cassette code tapes are not useful for learning the code.

95. Question-and-answer guides are available for all the amateur radio exams.

96. Learning International Morse Code is not practical for grade-school children.

97. A 100-W bulb has less resistance than a 40-W bulb.

98. GL means "go left."

99. ES means "and."

100. DE means "deliver message."

101. 73 means "best regards."

## Quiz Answers

1. True. All amateur radio operators must be licensed.

2. False. Although citizenship is required, the novice license also requires a code and written test.

3. False. Remember the "World's Youngest Ham"? Age 4, he was.

4. False. They are given by qualified radio amateurs.

5. False. These days, only about one out of five of us is an electronics specialist.

6. False. The FCC was established in 1934.

7. True. The purpose of the FCC is to regulate all radio broadcasting, of which amateur radio is one part.

8. False. The FCC issues all licenses in amateur radio.

9. True. Many amateurs are active in television broadcasting as a hobby. There are even TV repeaters around the country.

10. True. Amateur satellite use is a very important part of amateur radio these days. Other countries have them too.

11. False. We share our amateur radio satellites.

12. False. The most popular DX band is 20 m; 15 m is also a very good DX band, as is 40 m late at night.

13. True. The FM broadcast band is 88 to 108 MHz, whereas the highest HF band is 28.0 to 29.7 MHz.

14. True. *Kilo* means one thousand; *mega* means one million.

15. True. It is 1.7 MHz wide; that's 1700 kHz, more than the entire AM broadcast band.

16. False. Ten meters is normally a daytime band.

17. False. No one is required to take part in any operating activity in ham radio. It is all voluntary.

18. True. There are hundreds of them in operation at all hours of the day and night and on all bands.

19. False. They can stay and gab if they want. None of the amateur frequencies are reserved for anyone or anything.

20. False. Chess sure is permitted, welcomed, and enjoyed.

21. False. WAS means "Worked All States" in the United States.

22. False. The WAS certificate is issued by the ARRL.

23. True. Officially, it means to acknowledge a transmission or a message, but it has also taken on the meaning of the cards we send each other to acknowledge a contact.

24. False. Put whatever else you want on them as long as they have what the ARRL needs for proof of contact such as date; time; band; mode of operation (code or voice); and signal report. Dozens of printers make these cards with all this already printed on them.

25. False. Maybe a few countries are touchy about it because of their own restrictions, but, in general, we all exchange cards.

26. True. Everything we do in amateur radio operating is optional.

27. False. Everyone in amateur radio must have their own call sign and license.

28. False. They begin with Ws and Ks, mostly, but some also begin with A or N.

29. True. Ten call districts for the continental United States, with Alaska, Hawaii, and Puerto Rico having their own special prefix letters.

30. False. In the old days, yes, but not anymore. You can keep your old call sign no matter where you move.

31. True. DX takes on other meanings in our ham language such as referring to a foreign country, but its best definition is "distance."

32. True. It is one of the most exciting certificates to get except, possibly, WAS: 100 countries for DXCC, 50 states for WAS.

33. True. Particularly right before dawn, when the ionosphere bends radio waves into the longest skip. That's when those far, far countries come in best.

34. False. They count if they are recognized by the ARRL. The purpose of DX-peditions is to make available another new country in which there may not have been any radio amateurs.

35. False. He has several more countries than are in existence, and that's because countries come and go but DXers just hang in there.

36. True. It's in every dictionary too, so it really isn't as technical as it might first sound.

37. True. Normally, only the higher frequencies above the HF band go right through the ionosphere; the HF signals are bounced back. That's why we are able to hear signals from other parts of the world on the HF band but not on VHF.

38. False. The ionosphere thins out at night and moves higher, to where radio waves bouncing off it at night are reflected back at a greater distance. During the day the ionosphere gets heavier and lower.

39. False. Sun spots surely do have a lot of effect on radio transmissions.

40. False. No one is required to operate or take part in any activity; it is all voluntary.

41. False. The purpose of Field Day is to demonstrate that we are ready with our emergency equipment for an emergency of just about any sort. Making a lot of contacts is the fun way we have of testing our equipment in the field.

42. True. It takes only one. Some have 100, some have 10 to 30, which is probably a typical number.

43. True. We aren't forced into any public services by our amateur operating, but, if the hobby as a group did not offer public services, we would never have survived this long with our many benefits and privileges.

44. False. How about all those blind members I've mentioned?

45. False. Senator Goldwater has been in the hobby for over a half century.

46. True. Not for just kids but grownups, too. You get to learn about those countries before you can work the obscure ones. For example, even though I once sailed off Madagascar, it wasn't until my son became active in DX that I learned they had changed their name to Malagasy Republic in 1958.

47. False. A lot of stations operate *maritime mobile,* as it is called, when operating a radio station at sea.

48. True. It was an electrifying moment for the world of the time.

49. False. We have about 25,000 women in the hobby. And they are not all "career women" unless you properly count being a homemaker as a career.

50. True. And we are sincerely unchauvinistic in our use of this long-used hobby term.

51. False. "Old man" is what that one means.

52. False. Their members are from all around the world.

53. True. Ex-YL; meaning, "former (unmarried) young lady"; "wife."

54. False. There are several hundred ham families in the country, some with eight or nine family members if you count the in-laws.

55. False. They are primarily used for casual conversation, but they sure do come in handy in a strange area when you are looking for directions.

56. True. The higher the better for greater range, such as those on top of the Empire State Building or on mountain peaks.

57. False. Almost all of them are open to anyone who comes by.

58. False. Most walkie-talkies have a switch to reduce their power to one-tenth of a watt, enough to reach repeaters 10 to 20 mi away. (CB sets run 5 W, by comparison, so you can see how little power it takes to have a powerful lot of fun on amateur radio repeaters.)

59. False. When skip conditions come in, anything gloriously exciting on the ham bands may take place. Some radio amateurs are trying for WAS on 2-m repeaters!

60. True. Sorry about that, but such is the decision of the FCC. You can inquire about a hotel, but you simply can't make the reservation through amateur radio communications. The inquiry is personal; the reservation action is commercial.

61. True, so far, although there is much discussion going on these days with regard to issuing a no-code license. If it ever comes about, it will likely be for a class of license which will require a lot of theory testing.

62. False. We have a group of amateurs who regularly run on at 30 to 40 wpm (my son is one). The average speed on the air is about 15 wpm; it is much less on novice bands.

63. False. Being *code happy* is to be in a state of code learning in which the mind translates everything the eye sees into code. When you reach that state of mind, you are reaching a level of skill in code whereby you will soon be able to pass the code test.

64. True. There is the telegrapher's Morse, and there are

also variations used by a few other countries. However, the International Morse Code is what you will encounter on the ham bands.

65. True. And also 5 wpm for the technician class license.

66. False. As mentioned previously.

67. True. Every 10 years.

68. True. It is up to your examiner to give either multiple-choice or essay questions.

69. False. Several novices are already DXCC. DX, for novices, is quite a lot of fun on 15 m and 40 m, plus 10 m when the band is open.

70. False. The top grade is the extra class license.

71. False. Novice licenses are issued after the examiner grades their written exams and sends the passing grade to the FCC. The FCC then issues the call sign.

72. True. About half of the questions are on rules, etc.

73. True.

74. False. The flow of electricity is *current,* which is measured in units of amperes. Remember: Water and current *flow*.

75. True. R for resistance, measured in ohms.

76. False. The resistance of a circuit is equal to the voltage (10) divided by the current. 10V/5A = 2 Ω.

77. True. Glass is insulating material. It does not allow the flow of electricity.

78. False. They are graded by your examiner.

79. False. It means T for tone is only 5, which is a poor report.

80. False. That beautiful old equipment is very useful on the ham bands, particularly for novices. As cheap as they are at our hamfests, when you grow out of an old set of transmitter and receiver separates, put them away in the attic. They'll become antiques in another 10 or 20 years and worth twice what you paid for them. Try to pack some extra tubes with them, as the tubes are in very short supply and getting shorter. An antique rig that comes with extra tubes will get an extra price.

81. False. A transceiver is a combined transmitter and receiver in one cabinet. Certain circuits are used in both transmit and receive, which makes them less costly to build. CB sets are transceivers, for example.

82. False. Novices have no voice-operating privileges, only code. Technicians can operate voice on 2 m, however.

83. False. Beam antennas take a good-sized amount of cash, what with the rotors and towers you need to go along with them. Wire antennas can be strung from tree to house, etc., and wire is cheap. Any kind of wire will do; it doesn't have to be anything special although there is wire made especially for antenna work.

84. False. Not until you get a higher class license would you ever need to concern yourself with any extra house wiring. In fact, I run the legal limit of 1 kW but still use a regular wall outlet.

85. True. You bet; in fact, we encourage you to pick up a soldering iron, maybe not so much to build a transmitter (they're complicated, these days) or a receiver (they're twice as complicated), but because little gadgets are fun to build, and you can always use one more little gadget around the ham station.

86. False. Amateur radio is the only service in which home-built equipment is not only authorized but encouraged. All other radio services must be type-approved by the FCC before they can be used on the air.

87. True. The 40-m novice band is from 7.100 to 7.150 MHz.

88. False. The *ARRL Handbook* is the hobby's bible of technical information on just about everything technical we get into. It is published annually.

89. True. Even with *wire wrap* tools for modern IC (integrated-circuit) chips, there is always need for a soldering iron in ham radio.

90. False. You get alternating current, ac, from the electric company. You get dc from batteries.

91. True. *CQ Magazine* and the ARRL have one.

92. False. The FCC doesn't conduct courses; individuals and nongovernment organizations do.

93. False. A *short circuit* means a direct connection across the voltage, causing a large current to flow.

94. False. They are one of the best (and low-cost) ways to learn the code.

95. True. They cover just about every area of study too.

Study guides are quite useful and recommended prior to taking any of the amateur license exams.

96. False. (No one gets this question wrong.) The kids don't have our hesitations. They jump right in and learn in a hurry. The answer then is to clear your mind for code practice and enjoy it. The more you enjoy it, the faster you will learn.

97. True. The 100-W bulb draws more current because its resistance is lower.

98. False. GL means "good luck."

99. True. ES does mean "and," which comes from the old telegrapher's habits of shortening common words so that they could send messages faster.

100. False. DE means "from."

101. True. And my 73s to you.

# · GLOSSARY ·

The amateur-radio-related definitions given in this glossary are less formal than those found in textbooks. The purpose of these informal definitions is to provide the reader with brief and readily understood explanations of these technical terms.

*Absorption belt* An area in the atmosphere where oxygen and water vapor absorb energy from a radio wave.

*ac* See Alternating current.

*Access* (a repeater) To successfully contact a repeater station.

*Alternating current* [ac] Electromotive force (voltage) which periodically reverses direction of polarity and current flow.

*Amateur service* A hobby of radio communicating and personal skill development; licensed and regulated in the United States by the Federal Communications Commission.

*Ampere* [A or amps] (am' peer) Standard unit of electric current; named after French physicist André Marie Ampere.

*Amplifier* (am' pli' fy' yer) Electronic device to increase power of an electronic curcuit such as the output stage of a radio transmitter.

*Analog*  (an' a log)  In electronics, any electronics theory or circuits in which the signals are primarily nondigital, i.e., voice, music, sine waves, etc.

*Antenna*  (an ten' uh)  Wires used to send or receive radio transmission; usually designed to operate on specific bands; aerial.

*Appliance operator*  Hobby term, refers to radio operators who have little understanding of the technical equipment they operate.

*AR*  Code abbreviation, "end of transmission."

*ARES*  Amateur Radio Emergency Service.

*Armchair copy*  Hobby term, refers to hearing another station with ease; strong signal received with little or no interference.

*ARRL*  Amateur Radio Relay League.

*Atmospheric conditions*  (at' muh sfeer ik)  In radio, refers to characteristics of the upper atmosphere which have to do with the reflection of radio waves back to earth. *Good atmospheric conditions* refers to a normal situation in which the upper atmosphere is stable and in which signals reflect down to earth with minimum losses.

*Atmospherics*  Disturbances in reception produced by natural electric discharges such as static.

*Audio equipment*  (ah dee' yoe)  Electronic devices which produce audible sound output, for example, stereo amplifiers, record players, and cassettes.

*Auto patch*  Automatic telephone-connecting device which feeds radio transmitter to telephone lines; usually refers to use with repeaters.

*A1 Operator's Club*  Award issued by ARRL for having demonstrated quality operating skills on the amateur bands.

*Bands*  Assigned groups of frequencies in radio spectrum.

*Base (station)*  Home radio station; term seldom used in amateur radio; more commonly used in CB.

*Battery*  (bat'er ree)  Source of dc electromotive force (voltage); made up of chemical compounds, in cells, which act on metal electrodes to produce an emf; may be either wet cells or dry cells.

**Beam (antenna)** Directional antenna, usually made of aluminum tubing, supported by pipe secured to mast.

**Big switch** Hobby term, refers to disconnecting power to entire radio station.

**Boat anchor** Hobby term, refers to larger, outdated amateur radio sets of the 1950–1960 period, or earlier.

**Break in** To enter an existing conversation on the air. Also used in code conversations whereby one's transceiver automatically falls back into receive status inbetween code characters, thereby permitting any other station to break in at any time, even between letters being sent.

**Breakdown (voltage)** Level of applied voltage which causes insulating material to conduct electricity.

**Broadcast** To transmit; to send radio signal.

**Broadcast radio** Radio receiver which tunes broadcast bands, either AM, FM, or both.

**BT** Code abbreviation, used to separate sentences in paragraph or to separate any thoughts in transmission.

**Cable** Electric wire used to connect pieces of equipment such as from transmitter to antenna or transmitter to microphone.

**Call** Amateur call sign.

**Call book** Compilation of all licensed radio amateurs; available in book for U.S. amateurs only or for all amateurs in other parts of the world.

**Call sign** Set of letters/number which together form distinctive assignment of identification for all radio amateurs throughout the world, each with his or her own.

**Capacitor** (ka pass' ih tor) Electronic part, designed to store energy.

**CARI** Chess & Amateur Radio International.

**CB** See Citizen Band.

**Channel** Single frequency for either receive or transmit. Since radio amateurs use *bands*—groups of frequencies—channel is seldom used in the hobby except with regard to repeaters (which operate on single frequencies).

***Check in*** To call in to an existing contact on the air, for example, to check into a traffic net.

***Chip*** Active electronic device typically used in computers; consists of several hundred or thousands of transistors and other elements all located in one package (chip).

***Circuit*** (ser' kut) Collection of electrical parts in an operating electronic/electric unit.

***Citizen band*** (CB) Personal radio service located below the 10-m band in the radio spectrum. License not required.

***Class (license)*** Amateur radio license level; novice, technician, general, advanced, and extra class.

***Code*** In amateur radio, International Morse Code; collection of short and long sounds separated by spaces in recognized arrangements.

***Code happy*** Hobby term, refers to code students in which concentration becomes keenly developed and code characters become reflex action so that students automatically translate into code just about every word they see.

***Code speed*** Rate at which code is sent or received in terms of words per minute in which, by definition, five letters is equal to one word, but one number or punctuation counts as two letters. Example: *tomorrow* is equal to two words; *in 1984* is also equal to two words.

***Communications link*** A group of radio stations joined on a common frequency for a specific purpose such as in support of an ongoing emergency situation, drill, or traffic session.

***Component*** (com poe' nent) Electric part such as resistor, capacitor, coil, relay, or switch.

***Computer*** Electronic computing unit.

***Computer-aided*** Equipment with built-in computer chip which performs certain functions such as memory of preset frequencies and automatic band scanning.

***Conditions*** Natural atmospherics which determine how effective radio bands are for communications; also implies whether interference exists, either from other stations or natural sources.

***Conductor*** Any element that readily permits the flow of electricity; electric (metallic) wires.

***Connector*** Part used by which to join electric cables.

*Contact*   Radio communication; conversation on the air.

*Copy*   To receive, or hear, the other station's transmission. See Solid copy.

*Cord*   Electric cable used to connect electric units such as between house power and light bulb in lamp.

*Coverage*   Geographical area over which any transmitter can be heard. Commonly used in reference to repeater range.

*CQ*   Code abbreviation, "general call to any station."

*CQ DX*   Code abbreviation, CQ call to any foreign country station.

*Crystal receiver*   Ancient receiver set used in times prior to development of vacuum tubes. Required no power source.

*Current*   Flow of electricity measured in units of amperes.

*CW*   Continuous wave; another term for *code*. *Wave* refers to radio wave being transmitted; *continuous* refers to single frequency of transmitter; thus CW is a continuous wave interrupted in short and long series with appropriate spaces between, corresponding to Morse code.

*Cycle*   Electric signal that repeats itself in two reversals, one positive and one negative, to form a complete period.

*Cyclical*   Consecutive collection of cycles of electric signals, for example, alternating current.

*dc*   See Direct current.

*DE*   Code abbreviation, "from."

*Deci*   Latin prefix meaning "one-tenth"; 0.1.

*Decibel*   [dB] (dess′ ih bell)   Unit of signal power as measured against a reference level; used to show changes in power levels such as power of received signal.

*Detector (crystal)*   See Crystal receiver.

*Diagram*   (die′ a gramm)   Drawing of electric circuit.

*Digital (electronics)*   (dij′ i tul)   Electronics theory or circuit in which signals are made up primarily of pulses, such as computers and digital displays.

*Digital display*   Electronic number display used in place of dial indicators for, typically, frequency to which set is tuned.

*Diode*   Electric part which permits current to flow in only one direction. Typical use is in power supplies, to convert ac to dc. See Power supply.

***Dipole*** Wire antenna whose total length is equal to one-half wavelength at frequency of operation; fed by cable from radio set to center of antenna.

***Direct Current*** [dc] Electromotive force that causes current to flow in only one direction; battery.

***Distress signal*** Emergency call; in code, sent as SOS (amateurs sometimes use QRR); on voice, "Mayday."

***Dit-dah*** Verbal representation of International Morse Code sounds; dit = dot; dah = dash.

***DX*** Code abbreviation, "distance." Usually refers to foreign country or far-distant station normally outside of range.

***DXCC*** DX Century Club.

***DX-pedition*** When radio amateurs set up a temporary amateur station in remote country.

***DX pileup*** Large gathering of amateur operators, usually on same frequency, each attempting to contact a DX station.

***Earphones*** Receiving device worn over the ears; miniature loudspeakers to hear signals from radio receiver or audio equipment.

***Electrical principles*** Basic technical theory of electrical or electronic concepts.

***Electromotive force*** [emf] (ih lek' troe moe tiv' forse) Source of voltage.

***Electromagnetic spectrum*** Complete range of all radio bands and frequencies.

***Electrons*** (ih lek' trons) Small atomic particles of electricity surrounding the atom; flow of electric current is a flow of electrons.

***Elements (beam)*** Supported aluminum tubing arranged in special lengths and spacing to form a directional antenna.

***Elmer*** Hobby term, refers to any licensed radio amateur who helps someone to join the hobby.

***Emergency preparedness*** Having one's radio station and/or self available for use in emergency communications.

***Emissions*** (ih mish' uns) Signals broadcast from radio transmitter.

***ES*** Code abbreviation, "and."

***Exam (amateur radio)*** Test of applicant's radio communications abilities in both code and theory for various classes of licenses.

*Examiner (volunteer)*   Qualified radio amateur who gives tests to amateur radio applicants.

*FCC*   Federal Communications Commission.

*Field Day*   Annual amateur operating activity sponsored by ARRL for purpose of demonstrating emergency preparedness of equipment and operators; conducted overnight, one weekend in June.

*Filter*   Electronic device typically used in power supplies, consisting of capacitors; used to smooth ripples of alternating current into pure direct current.

*FM broadcast band*   Entertainment broadcast band, 88 to 108 MHz, using voice modulation technique known as "frequency modulation."

*Ford coil*   Ignition coil from ancient Ford cars; popularly used in spark-gap transmitters from early days of amateur radio; coil increases voltage which causes spark to develop.

*Frequency*   The number of hertz [Hz] (cycles) of an alternating current.

*Fuse*   Metal strip which limits maximum current flow in a circuit; melts when design current is exceeded, creating an open circuit, after which no current will flow, thereby protecting other circuit elements.

*Gear*   Hobby term, refers to radio equipment.

*Generator*   Electric power unit, often powered by gasoline, diesel fuel, or batteries.

*GL*   Code term, "good luck."

*GM*   Code term, "good morning."

*Guy wire*   Cable used to support towers; anchored on one end to points on ground.

*Ham*   Popular name for amateur radio operator.

*Hamdom*   Hobby term, refers to all of ham radio.

*Hamfest*   Ham radio flea market, usually outdoors, where ham radio equipment—commercial or home built, new or used—is displayed for sale.

***Handi-Ham*** Amateur radio organization dedicated to helping the handicapped become amateur radio operators.

***Hardware*** In electronics, refers to physical electronic equipment. See Software.

***Harmonic*** Hobby term, one's offspring. Also, technical term to mean a frequency that is an exact multiple of another. Example: 14.0 MHz is the second harmonic of 7.0 MHz.

***Hertz*** [Hz] Unit of frequency, cycle; named after German physicist Heinrich Hertz.

***HF bands*** High frequency bands; officially those bands between the frequencies of 3.0 and 30.0 MHz.

***High technology*** Advanced developments in theory and circuitry such as in electronic computers.

***Home brew*** Hobby term, home-built equipment; non-commercial.

***House current*** Power supplied to homes from electric company, available at electric outlets.

***HT*** Abbreviation for handie-talkie; walkie-talkie.

***IARU*** International Amateur Radio Union.

***IMRA*** International Mission Radio Association.

***Incentive licensing*** Refers to past changes in amateur licensing which created the current various classes of licenses.

***Insulator*** Material which normally does not conduct electricity.

***Interference*** Disturbances to reception of radio signals either from normal atmospheric conditions or from other stations on same or nearby frequency; also, from electric devices such as electric motors and thermostats.

***Ionosphere*** (eye on' esfir) Outer part of earth's atmosphere from which radio waves are reflected back to earth.

***ITU*** International Telecommunications Union.

***Jumper (cable)*** External piece of wire connected across part of an electric circuit or electric device.

***K*** Code abbreviation, invitation to transmit.

***Key (code)*** Device used by code operator to manually form code characters.

*Kilo*  [k] (kil′ oh)  Greek prefix, one thousand (1000).
*Kilohertz*  [kHz]  1000 hertz.

*License*  Certificate presented to applicants who successfully complete an examination on amateur radio concepts.
*Light-year*  In astronomy, the distance traveled by light (or radio waves) in one year; about 6 trillion mi.
*Line of sight*  Straight-line propagation path between two radio stations. VHF radio waves travel line of sight; HF waves travel up to atmosphere and are reflected back down.
*Log*  Record of radio contacts. Not required in amateur radio; maintained by many operators for personal benefit.
*Logging*  To record one's contacts in certain detail such as station contacted, date, time, frequency, and signal report.
*Loudspeaker*  Electromechanical device which converts electric signals into audible signals; performs reverse function of microphone. See Microphone.
*Low pass filter*  Device which passes low frequencies while suppressing higher frequencies. Example: Filter connected to output of transmitter passes lower HF frequencies (amateur bands) while suppressing unintentional higher frequencies (VHF). Used to reduce interference from amateur sets to television sets.

*Magnetic base antenna*  A mobile antenna with large magnet in base permitting simple and temporary attachment to metal car roof.
*Magnetic tape*  Electronic storage device used in computers.
*MARS*  Military Affiliate Radio Service.
*Mega* [M]  Greek prefix, one million (1,000,000).
*Megahertz* [MHz]  1,000,000 hertz.
*Memory circuit*  Computer chip built into radio set which permits presetting fixed numbers/channels for single-button recall.
*Message*  Information originated, relayed, or received via radio.
*Micro* [μ]  Millionth part of a unit (0.000001).
*Microcomputer*  Small computer.
*Microphone*  Electromechanical device which converts

voice into the electric signals which are applied to a radio transmitter, audio amplifier, public address system, etc.

*Milli* [m]   Latin prefix, one-thousandth (0.001).

*Minus*   Negative voltage such as negative polarity ( − ) of battery; negative half of ac cycle.

*Mobile (radio)*   Any moving radio station in land service. (Others: *aeronautical mobile,* station on aircraft; *maritime mobile,* station on board ship.)

*Mode (radio transmission)*   Type of transmission used such as voice, code, TV, or radioteletype.

*Morse code*   See International Morse Code.

*Multiple-choice (questions)*   Test questions which have either four or five answers from which to select the correct one.

*Murphy's law*   Collection of informal humorous "laws" which seem to govern our lives in contradicting ways. Example: Murphy's law says that when everything is working perfectly, that is the time to beware; something must go wrong. Or, when the band is otherwise completely dead, there will be one station on the air—a station that continuously tests without identifying, and they will be coming in very, very strongly. Or, when you are about to call a friend on a schedule and the frequency has been absolutely clear for the past half hour, the instant you pick up the microphone to transmit, someone with an overpowering signal will have just begun a long CQ call.

*Negative*   See Minus.

*Net*   Network of radio stations arranged for a common purpose such as handling/relaying traffic messages.

*Net control*   Station in charge of net.

*Ohm* [Ω]   Unit of electrical resistance, named after German physicist Georg Ohm.

*Ohm's law*   Electrical concept developed by Ohm; resistance in circuit is proportional to voltage applied and inversely proportional to current in the circuit.

*OM*   Code abbreviation, "old man."

*One-eyed monster*   Hobby term, refers to television interference, TVI, in which the I in *interference* is the one-eyed monster.

***Open circuit*** Electric circuit in which there is no continuity between one terminal of the voltage source (emf) and the other; hence no current can flow.

***Operating procedures (amateur)*** Methods established by law or by common agreement for most effective radio communications.

***OSCAR*** Orbitting Satellite Carrying Amateur Radio.

***Phone*** Radio communications by voice.

***Phone patch*** Connection of radio transceiver to telephone line.

***Phonetics*** Word substitutions for each letter of alphabet; used to assure effective exchange of information over the air. Example: K2VJ, "Kilo Two Victor Juliette."

***Pico*** [p] One trillionth (0.000000001).

***Plate*** Output circuit of vacuum tube circuit.

***Plate transformer*** Device used to step up ac voltage to high levels required for plate circuit of vacuum tubes. See Vacuum tubes; Transformer.

***Plus*** See Positive.

***Polarity*** Direction or level of electric signal with respect to zero; either positive (plus, $+$) or negative (minus, $-$).

***Pole*** Terminal of battery (dc); either positive or negative terminals.

***Positive*** Plus polarity of electric signal.

***Power*** Product of voltage and current in electric circuit. Expressed in units of watts (W) (named after Scottish scientist James Watt).

***Power supply*** Electric unit used to convert primary voltage from alternating current to direct current; from direct current to alternating current; or direct current to direct current. Output voltage may be higher than, or lower than, input voltage, as required. Voltage changes are obtained by transformer and conversion from ac to dc by rectifier; output of rectifier is smoothed to pure dc by filters.

***Practices (radio)*** See Operating practices.

***Prefix (call sign)*** Portion of assigned call sign containing one or two letters before the number of the call sign district.

***Preparedness*** See Emergency preparedness.

*Priority*   Radio communications situation of more urgent nature than routine.

*Propagation*   (prop' a gay shun)   Path of radio wave through atmosphere.

*Q&A Manual*   Question and answer manuals on radio exams.

*QRM*   International Q signal: "I am being interfered with."

*Q.R. Zedd*   Legendary fiction character noted for unsurpassed skills at DX; written in *Collector & Emitter Newsletter*.

*Q Signals*   Abbreviations internationally recognized in radio communications; permits operator from any country to exchange certain information with operator from any other country without knowing each other's language.

*QSL*   International Q signal: "I acknowledge your transmission." Also refers to cards exchanged after an amateur contact, i.e., QSL cards.

*QSL Bureau*   ARRL division which processes QSL cards exchanged between DX and U.S. stations.

*QSO*   International Q signal: "I am in contact with."

*QST*   Official publication of the ARRL. Also refers to transmission of a bulletin to any and all stations.

*R*   Code abbreviation, "roger"; also, R for "readability" in signal report given over the air.

*Radiate*   To broadcast a radio signal.

*Radio frequency*   Single-spot location in radio spectrum; measured in kilohertz (kHz) or megahertz (MHz).

*Radio set*   Amateur radio transmitter, receiver, or transceiver.

*Radio spectrum*   See Electromagnetic spectrum.

*Radiogram*   Message sent by radio.

*Radioteletype*   Teletype signals sent by radio.

*Rag chew*   Informal conversation over the air.

*RCC*   Rag Chewer's Club.

*Receiver*   Equipment used to detect radio signals.

*Reception*   Receiving conditions on the bands. Poor reception may be caused by atmospheric instabilities or by interference from other stations.

*Rectifier*   Circuit which converts ac to dc; consists of diodes; common expression for *power supply*.

*Regeneration*   Technique of feeding back signals from output to input to greatly increase gain or sensitivity of a circuit.

*Regulator (power)*   Transformer which automatically provides a stabilized output voltage regardless of variations in input voltage.

*Relay messages*   Practice of sending messages across great distances by handing off from one radio operator to the next.

*Relay rack*   Metal shelving with panels and enclosure; used to house large amounts of electronic equipment; commonly used in the past when radio transmitters were physically large.

*Repeater*   Remote station, usually in VHF bands, used to automatically relay amateur transmissions.

*Rhombic (antenna)*   Lengthy wire antenna, diamond shaped, often running thousands of feet on each leg. Offers very great sensitivity to reception and transmission of radio signals.

*Rig*   Amateur radio set.

*Roll your own*   Hobby term, to build one's own equipment.

*Rolling in*   Hobby term, refers to reception of signals heard rather loudly.

*Rotor (beam)*   Motorized device used to rotate beam antenna.

*RST*   Code abbreviation; *R*eadability, *S*ignal strength and *T*one; used in exchanging signal reports on the amateur bands.

*RTTY*   Radio teletype. See Teletype.

*S*   Code abbreviation, *S*ignal strength.

*Satellite communications (amateur)*   Radio relay via space satellite under amateur radio control.

*Scanner*   Computer-aided circuitry which provides automatic sweep tuning up and down a preselected range of radio frequencies in a receiver.

*Schedule*   Prearranged time and frequency for meeting on the air.

*Send*   To transmit or broadcast.

*Separates*   Hobby term, refers to radio set consisting of separate transmitter and receiver as opposed to transceiver which has both units combined in one package.

*Servicing*   To repair or maintain equipment.

*Set*   Radio equipment.

*Shack*   Hobby term, room in which amateur radio station is located.

*Shock*   Result of physical contact with electric power of sufficiently high level as to cause unpleasant or harmful physical reaction.

*Shock hazard*   Any improperly exposed point of voltage.

*Short circuit*   Direct connection across any voltage source such as to cause large and uncontrolled current flow.

*Shortwave*   Radio frequencies in the HF range (3.0 to 30.0 MHz). Often referred to as "HF" or "high frequencies."

*Signal*   Refers to received broadcast.

*Signal report*   Exchange of reception quality in terms of readability, signal strength, and tone quality. See RST.

*Sine wave*   Cycle of alternating voltage or current.

*Skip*   Reflection or bounce of radio signal from atmosphere.

*Sloper (antenna)*   Type of antenna hung at an angle from the vertical.

*S meter*   Voltmeter internally connected to receiver to measure strength of incoming signals; calibrated from 0 to 9, and in decibels over 9.

*Software*   Computer programs/instructions.

*Solar radiation*   Intense solar eruptions which cause sudden increases in absorption of the ionosphere to radio signals; shows up as partial or complete radio blackout.

*Solder*   Low-melting-temperature metal used to bond electric wires.

*Soldering iron*   Electric device used to heat solder and wires being bonded by solder.

*Solid copy*   Hobby term, refers to received signals which are clearly heard. See Armchair copy.

*SOS*   See Distress signal.

*Spark coil*   Ignition coil (see Ford coil) used in ancient spark-gap transmitters.

*Spark gap*   Refers to gap in spark-gap transmitters across which the high voltage would create an arc, the arc radiation being applied to a tuned circuit which determined the frequency of operation of the transmitter.

**Sparks**   Title given to shipboard radio operator.

**Spectrum**   See Electromagnetic spectrum.

**SSB**   Single sideband; modern form of voice transmission; uses half the spectrum space of older voice transmitters.

**State of the art**   Most recent developments in scientific matters.

**Station**   Amateur radio facilities; includes at least a transmitter, receiver, antenna, and key or microphone; shack.

**Sun spots**   Active dark spots on surface of sun which result in disturbances to radio communications; occurs in cycles of 11-year periods.

**Switch**   Device used to open or close an electric circuit.

**T**   Code abbreviation; *T*one.

**Telegram**   Message sent in code by wire telegraphy.

**Telephone access**   See Auto patch.

**Teletype**   Electric typewriting device which is used to send and receive special code at high speed.

**Terminal**   In batteries, refers to one end, either positive or negative; also called *pole.*

**Theory**   Explanations of technical subjects.

**Ticket**   Hobby term, refers to amateur license.

**Tinkerer**   In amateur radio, one who is fond of constructing or modifying electronic units.

**TMI**   Three Mile Island nuclear power plant near Harrisburg, Pennsylvania.

**Touch-tone pad**   Telephone touch dial in amateur transmitter which permits dialing telephones in repeaters which are connected to telephone lines via auto patch.

**Traffic**   Messages sent by radio.

**Transceiver**   Combination transmitter and receiver built in one cabinet; uses certain common circuitry.

**Transformer**   Electromechanical device used with ac voltage to step up or down. Commonly used in power supplies.

**Transmitter**   Radio set used to broadcast over the air.

**Tune**   To properly adjust a radio set.

**TVI**   Television interference.

***Upgrade*** To move to a higher class of amateur license.

***Vacuum tube*** Sealed glass device containing elements within vacuum; permits amplification of very low signal levels.
***VHF*** Very high frequencies, formally in the range of 30.0 to 300.0 MHz.
***Volt*** [V] Unit of voltage.
***Voltage*** Electromotive force.
***Volunteer examiner*** See Examiner.

***W1AW*** Headquarters amateur radio station of ARRL; call sign of ARRL cofounder Hiram Percy Maxim.
***W4-station*** Hobby term, refers to any amateur radio station in fourth call sign district.
***WAC*** Worked All Continents.
***Walkie-talkie*** See HT.
***WAS*** Worked All States.
***Watt*** [W] Unit of power. See Power.
***Wavelength*** Distance between corresponding points of two successive radio waves; measured in meters.
***Wire wrap*** Tool used to connect digital electronics components such as IC chips; wires are wrapped around very small terminals of chips rather than being soldered to the terminals.
***Wireless*** Communication without using wires interconnecting transmitting and receiving stations.
***Wires*** Conductors which are used to carry electric signals.
***Worked*** Hobby term, contacted.
***Worldwide DX contest*** Sponsored by *CQ Magazine;* held annually; points are acquired for numbers of stations contacted in foreign countries, plus other features.
***WPM*** Words per minute.

***XYL*** Hobby term, wife; ex Young Lady; ex-YL.

***YL*** Hobby term, Young Lady.
***YLCC*** Award given for contacting YLs in 100 different countries.
***YLRL*** Young Ladies' Relay League.

# · Appendix 1 ·

# Other Addresses
# of Interest

Amateur Radio News Service
528 Montana
Holton, KS 66436

American Shortwave Listener's Club
16182 Ballad Lane
Huntington Beach, CA 92649

auto-call
121 South Highland Street
Arlington, VA 22204

Buckmaster Publishing
70 Florida Hill Road
Ridgefield, CT 06877

Chess & Amateur Radio International
Post Office Box 682
Cologne, NJ 08213

Collector & Emitter
1020 Arthur Drive
Midwest City, OK 73110

CQ Magazine/Bookshop
76 North Broadway
Hicksville, NY 11801

Ham Radio Magazine/Book Store
Greenville, NH 03048

International DX'ers Club of San Diego
1826 Cypress
San Diego, CA 92154

International Mission Radio Association
Bro. Bernard Frey, O.F.M., WA2IPM
One Pryer Manor Road
Larchmont, NY 10538

Junior High School 22 ARC
Office of the Principal
111 Columbia Street
New York, NY 10002

Montebello Elementary School
Office of the Principal
Suffern, NY 10901

Quarter Century Wireless Association
1409 Cooper Drive
Irving, TX 75061

Radio Amateur Callbook, Inc.
925 Sherwood Drive
Lake Bluff, IL 66044

Radio Amateur Satellite Corp.
Post Office Box 27
Washington, DC 20044

Radio Shack
Post Office Box 2625
Fort Worth, TX 76101

Rocco Laurie Intermediate School 72
Office of the Principal
33 Ferndale Avenue
Staten Island, NY 10314

73 Magazine/Bookstore
Peterborough, NH 03458

W5YI Report
Post Office Box 10101
Dallas, TX 75207

WorldRadio, Inc.
2120 28th Street
Sacramento, CA 95819

Young Ladies Relay League
Minerva Fronhofer, WB2JNL
RFD 1
Salem, NY 12865

# · Appendix 2 ·

# FCC Field Offices

ALASKA:
　　1011 East Tudor Road, Room 240
　　Anchorage, AL 99510
　　Telephone: (907) 563–3899

CALIFORNIA (La Mesa):
　　7840 El Cajon Boulevard, Room 405
　　La Mesa, CA 92041
　　Telephone: (619) 293–5478

CALIFORNIA (Long Beach):
　　3711 Long Beach Boulevard, Room 501
　　Long Beach, CA 90807
　　Telephone: (213) 426–4451

CALIFORNIA (San Francisco):
　　423 Customhouse
　　555 Battery Street
　　San Francisco, CA 94111
　　Telephone: (451) 556–7701

COLORADO:
　　12477 West Cedar Drive
　　Denver, CO 80228
　　Telephone: (303) 234–6977

FLORIDA (Miami):
8675 Northwest Fifty-third Street, Suite 203
Miami, FL 33166
Telephone: (305) 350–5542

FLORIDA (Tampa):
Interstate Building, Room 601
1211 North Westshore Boulevard
Tampa, FL 33607
Telephone: (813) 228–2872

GEORGIA:
Room 440, Massell Building
1365 Peachtree Street, NE
Atlanta, GA 30309
Telephone: (404) 881–3084

HAWAII:
Prince Kuhio Federal Building
300 Ala Moana Boulevard, Room 7304
Post Office Box 50023
Honolulu, HI 96850
Telephone: (808) 546–5640

ILLINOIS:
230 South Dearborn Street, Room 3940
Chicago, IL 60604
Telephone: (312) 353–0195

LOUISIANA:
1009 F. Edward Hebert Federal Building
600 South Street
New Orleans, LA 70130
Telephone: (504) 589–2095

MARYLAND:
George M. Fallon Federal Building
Room 1017
31 Hopkins Place
Baltimore, MD 21201
Telephone: (301) 962–2728

MASSACHUSETTS:
1600 Customhouse
165 State Street
Boston, MA 02109
Telephone: (617) 223–6609

MICHIGAN:
> 1054 Federal Building and U.S. Courthouse
> 231 West LaFayette Street
> Detroit, MI 48226
> Telephone: (313) 226–6078

MINNESOTA:
> 691 Federal Building
> 316 North Robert Street
> Saint Paul, MN 55101
> Telephone: (612) 725–7810

MISSOURI:
> Brywood Office Tower
> Room 320, 800 East Sixty-third Street
> Kansas City, MO 64133
> Telephone: (816) 926–5111

NEW YORK (Buffalo):
> 1307 Federal Building
> 111 West Huron Street
> Buffalo, NY 14202
> Telephone: (716) 846–4511

NEW YORK (New York):
> 201 Varick Street
> New York, NY 10014
> Telephone: (212) 620–3437

OREGON:
> 1782 Federal Office Building
> 1220 Southwest Third Avenue
> Portland, OR 97204
> Telephone: (503) 221–4114

PENNSYLVANIA:
> One Oxford Valley Office Building
> 2300 East Lincoln Highway, Room 404
> Langhorne, PA 19047
> Telephone: (215) 752–1324

PUERTO RICO:
> Federal Building and Courthouse
> Room 747 Avenida Carlos Chardon
> Hato Rey, PR 00918
> Telephone: (809) 753–4567

TEXAS (Dallas):
Earle Cabell Federal Building
Room 13E7, 1100 Commerce Street
Dallas, TX 75242
Telephone: (214) 767–0761

TEXAS (Houston):
5636 Federal Building
515 Rusk Avenue
Houston, TX 77002
Telephone: (713) 229–2748

VIRGINIA:
Military Circle
870 North Military Highway
Norfolk, VA 23502
Telephone: (804) 441–6472

WASHINGTON:
3256 Federal Building
915 Second Avenue
Seattle, WA 98174
Telephone: (206) 442–7653

# · Appendix 3 ·

# International Amateur Radio Union Addresses

The following list of member societies to the International Amateur Radio Union is given for two reasons:

1. So that readers from other countries who have had their interest in amateur radio aroused by this book will have an address with which to contact someone in their home country for further information on their own amateur radio regulations and requirements

2. So that readers can see for themselves from such a substantial display of societies, that radio amateurs are indeed international in their operations and organization.

ALGERIA:

Amateurs Radio Algeriens (ARA)
Post Office Box 2
Algers, Algeria
Telephone: 64.57.07

ANDORRA:

Unio Radioaficionats Andorrans (URA)
Post Office Box 150
La Vella, Andorra
Telephone: 20160

ARGENTINA:
Radio Club Argentino (RCA)
Carlos Calvo 1420/24
Buenos Aires, Argentina 1102
Telephone: 26–0505

AUSTRALIA:
Wireless Institute of Australia (WIA)
Post Office Box 150
Toorak, Victoria 3141 Australia
Telephone: (03) 528 5962

AUSTRIA:
Osterreichischer Versuchssenderverband (OVSV)
Post Office Box 999
A-1014 Vienna, Austria
Telephone: 0222–632402

BAHAMAS:
Bahamas Amateur Radio Society (BARS)
Post Office Box 6249
Nassau, Bahamas

BAHRAIN:
Amateur Radio Association Bahrain (ARAB)
Post Office Box 22381
Muharraq, Bahrain
Telephone: 713937

BANGLADESH:
Bangladesh Amateur Radio League (BARL)
G.P.O. Box 3512
Dhaka, Bangladesh
Telephone: 316100

BARBADOS:
Amateur Radio Society of Barbados (ARSB)
Post Office Box 814E
Bridgetown, Barbados
Telephone: 809–426–2052

BELGIUM:
Union Belge des Amateurs-Emetteurs (UBA)
c/o Rene A. Vanmuysen, ON4VY, Diepestraat 54
1970 Wezembeek-Oppem, Belgium
Telephone: 02–731.42.86

BERMUDA:
Radio Society of Bermuda (RSB)
Post Office Box 275
Hamilton 5, Bermuda
Telephone: 809–298–2914

BOLIVIA:
Radio Club Boliviano (RCB)
Post Office Box 2111
La Paz, Bolivia

BOTSWANA:
Botswana Amateur Radio Society (BARS)
Post Office Box 1873
Gaborone, Botswana
Telephone: 53010

BRAZIL:
Liga de Amadores Brasileiros de Radio Emissao (LABRE)
Post Office Box 07–0004
70.200 Brasilia, D.F., Brazil
Telephone: (061) 223–1157

BRITISH VIRGIN ISLANDS:
British Virgin Islands Radio League (BVIRL)
Post Office Box 4
West End, Tortola, British Virgin Islands

BULGARIA:
Bulgaria Federation of Radio Amateurs (BFRA)
Post Office Box 830
Sofia, Bulgaria
Telephone: 87 21 13

BURMA:
Burma Amateur Radio Transmitting Society (BARTS)

CANADA:
Canadian Radio Relay League (CRRL)
Post Office Box 7009, Station E
London, Ontario N5Y 4J9 Canada
Telephone: 416–494–8721

CAYMAN ISLANDS:
Cayman Radio Society (CRS)
Post Office Box 1215
Grand Cayman, Cayman Islands
Telephone: 9–5922

CHILE:
>
> Radio Club de Chile (RCC)
> Post Office Box 13630
> Santiago, Chile
> Telephone: 647070724049

COLOMBIA:
>
> Liga Colombiana de Radioaficionados (LCRA)
> Post Office Box 584
> Bogotá, Colombia
> Telephone: 92457536

COSTA RICA:
>
> Radio Club de Costa Rica (RCCR)
> Post Office Box 2412
> San Jose, Costa Rica
> Telephone: 21–69–03

CUBA:
>
> Federacion de Radioaficionados de Cuba (FRC)
> Post Office Box 1
> Habana 1 Cuba
> Telephone: 302223

CYPRUS:
>
> Cyprus Amateur Radio Society (CARS)
> Post Office Box 1267
> Limassol, Cyprus
> Telephone: 051–65087

CZECHOSLOVAKIA:
>
> Central Radio Club of Czechoslovakia (CRCC)
> Post Office Box 69
> 113 27 Praha 1, Czechoslovakia

DENMARK:
>
> Eksperimenterende Danske Radioamatorer (EDR)
> Post Office Box 79
> DK-1003 Copenhagen K, Denmark
> Telephone 08–942224

DJIBOUTI:
>
> Association des Radioamateurs de Djibouti (ARAD)
> Post Office Box 1076
> Djibouti

**DOMINICAN REPUBLIC:**
Radio Club Dominicano (RCD)
Post Office Box 1157
Santo Domingo, Dominican Republic
Telephone: 533–2211

**ECUADOR:**
Guayaquil Radio Club (GRC)
Post Office Box 5757
Guayaquil, Ecuador
Telephone: 011–593–392–671

**EL SALVADOR:**
Club de Radioaficionados de El Salvador (CRAS)
Post Office Box 517
San Salvador, El Salvador
Telephone: 22–6491

**FAROE ISLANDS:**
Faroes Radio Amateurs (FRA)
Post Office Box 343
DK-3800 Torshavn
Faroe Islands via Denmark
Telephone: 1 26 39

**FIJI:**
Fiji Association of Radio Amateurs (FARA)
Post Office Box 184
Suva, Fiji
Telephone: 381605

**FINLAND:**
Suomen Radioamatooriliitto (SRAL)
Post Office Box 306
SF-00101
Helsinki 10, Finland
Telephone: 90–656109

**FRANCE:**
Reseau des Emetteurs Francais (REF)
Two Square Trudaine
75009 Paris, France
Telephone: 1 526 55 44

GAMBIA:

Radio Society of the Gambia (RSTG)
c/o K. Bone, C5ABK, MRC
Post Office Box 273
Banjul, Gambia
Telephone: Serre Kunda 2326

GERMANY:

(Democratic Republic of):
Radio klub der D.D.R. (RKDDR)
Langenbeckstrasse 36–39
1272 Neuenhagen b
Berlin, Germany
Telephone: 890

GERMANY:

(Federal Republic of):
Deutscher Amateur-Radio-Club (DARC)
Post Office Box 1155
D-3507 Baunatal, Germany
Telephone: 0561–492004

GHANA:

Ghana Amateur Radio Society (GARS)
Post Office Box 3773
Accra, Ghana

GIBRALTAR:

Gibraltar Amateur Radio Society (GARS)
Post Office Box 292
Gibraltar

GREECE:

Radio Amateur Association of Greece (RAAG)
Post Office Box 564
Athens, Greece
Telephone: (01) 5228175

GRENADA:

Grenada Amateur Radio Club (GARC)
Post Office Box 290
Saint George's, Grenada
Telephone: 2954/3141

GUATEMALA:

Club de Radioaficionados de Guatemala (CRAG)
Post Office Box 115
Guatemala City, Guatemala
Telephone: 514658

GUYANA:

Guyana Amateur Radio Association (GARA)
c/o Guyana Telecommunications Corp.
55 Brickdam
Georgetown, Guyana

HAITI:

Radio Club d'Haiti (RCH)
Post Office Box 1484
Port-au-Prince, Haiti

HONDURAS:

Radio Club de Honduras (RCH)
Post Office Box 273
San Pedro Sula, Honduras

HONG KONG:

Hong Kong Amateur Radio Transmitting Society (HARTS)
Post Office Box 541
Hong Kong
Telephone: 5–920091

HUNGARY:

Magyar Radioamator Szovetseg (MRASZ)
Post Office Box 214
H-1368 Budapest, Hungary

ICELAND:

Islenzkir Radioamatorar (IRA)
Post Office Box 1058
121 Reykjavik, Iceland

INDIA:

Amateur Radio Society of India (ARSI)
Post Office Box 3005
New Delhi 3, India
Telephone: 61 17 09

INDONESIA:

Organisasi Amatir Radio Indonesia (ORARI)
Post Office Box 97
Kbyt
Jakarta Selatan, Indonesia
Telephone: 761674

IRELAND:

Irish Radio Transmitters Society (IRTS)
Post Office Box 462
Dublin 9, Ireland

ISRAEL:

Israel Amateur Radio Club (IARC)
Post Office Box 4099
Tel-Aviv 61040 Israel
Telephone: 04–256064

ITALY:

Associazione Radioamatori Italiani (ARI)
Via Scarlatti 31
Milano 20124, Italy
Telephone: 02–202894

IVORY COAST:

Association des Radio-Amateurs Ivoiriens (ARAI)
Post Office Box 2946
Abidjan 01, Ivory Coast

JAMAICA:

Jamaica Amateur Radio Association (JARA)
76 Arnold Road
Kingston 5, Jamaica
Telephone: (809) 926–7246

JAPAN:

Japan Amateur Radio League (JARL)
1-14-2 Sugamo, Toshima
Tokyo 170 Japan
Telephone: (03) 947–8221

JORDAN:

Royal Jordanian Radio Amateur Society (RJRAS)
Post Office Box 2353
Amman, Jordan
Telephone: 666235

KENYA:

Radio Society of Kenya (RSK)
Post Office Box 45681
Nairobi, Kenya

KOREA:

Korean Amateur Radio League (KARL)
C.P.O. Box 162
Seoul 100, Korea
Telephone: 74–6770

LEBANON:

Association des Radio-Amateurs Libanais (RAL)
Post Office Box 8888
Beirut, Lebanon
Telephone: 241698

LIBERIA:

Liberia Radio Amateur Association (LRAA)
Post Office Box 1477
Monrovia, Liberia
Telephone: 223961

LUXEMBOURG:

Reseau Luxembourguida des Amateurs d'Ondes Courtes (RL)
c/o Josy Kirsch
23 Route de Noertzange
L-3530 Dudelange, Luxembourg
Telephone: 51 51 51

MALAYSIA:

Malaysian Amateur Radio Transmitters' Society (MARTS)
Post Office Box 777
Kuala Lumpur, Malaysia
Telephone: 03–565581

MALTA:

Malta Amateur Radio League (MARL)
Post Office Box 575
Valletta, Malta

MAURITIUS:

Mauritius Amateur Radio Society (MARS)

MEXICO:

Liga Mexicana de Radio Experimentadores (LMRE)
Post Office Box 907
06000 Mexico, D.F.
Telephone: 563–1405

MONACO:

Association des Radio-Amateurs de Monaco (ARM)
Villa Marie-Joseph
24 Av. Prince Pierre 98000, Monaco
Telephone: (33–93) 50.75.05

MONTSERRAT:

Montserrat Amateur Radio Society (MARS)
Post Office Box 448
Plymouth, Montserrat

MOROCCO:

Association Royale des Radio-Amateurs du Maroc (ARRAM)
c/o M'Rabety Driss, CN8BH
3 Rue Al-Farabi
Rabat, Morocco

MOZAMBIQUE:

Liga dos Radio Emissores de Mocambique (LREM)

NETHERLANDS:

Vereniging voor Exp. Radio Onderzoek in Nederland (VERON)
Post Office Box 1166
6801 BD Arnhem
Telephone: 085–426760

NETHERLANDS ANTILLES:

Vereniging voor Exp. Radio Onderzoek in Ned. Antillen
(VERON)
Post Office Box 3383
Curacao, Netherlands Antilles
Telephone: 614525

NEW ZEALAND:

New Zealand Association of Radio Transmitters (NZART)
Post Office Box 40–525
Upper Hutt, New Zealand
Telephone: Wellington 282170

NICARAGUA:

Club de Radioexperimentadores de Nicaragua (CREN)
Post Office Box 925
Managua, Nicaragua
Telephone: 71274

NIGERIA:

Nigerian Amateur Radio Society (NARS)
Post Office Box 2873
Lagos, Nigeria
Telephone: 870269

NORWAY:

Norsk Radio Relae Liga (NRRL)
Post Office Box 21
Refstad
Oslo 5, Norway
Telephone: (02) 22 51 86

OMAN:

Royal Omani Amateur Radio Society (ROARS)
Post Office Box 981
Muscat, Oman

PAKISTAN:

Pakistan Amateur Radio Society (PARS)
Post Office Box 65
Lahore, Pakistan
Telephone: 306380

PANAMA:

Liga Panamena de Radio Aficionados (LPRA)
Post Office Box 175
Panama 9a

PAPUA NEW GUINEA:

Papua New Guinea Amateur Radio Society (PNGARS)
Post Office Box 204
Port Moresby, Papua New Guinea
Telephone: 256933

PARAGUAY:

Radio Club Paraguayo (RCP)
Post Office Box 512
Asuncion, Paraguay
Telephone: 46–124

PERU:

Radio Club Peruano (RCP)
Post Office Box 538
Lima, Peru
Telephone: 414837

PHILIPPINES:

Philippine Amateur Radio Association (PARA)
Post Office Box 4083
Manila, Philippines
Telephone: 77–27–64

POLAND:

Polski Zwiazek Krotkofalowcow (PZK)
Post Office Box 320
00–95 Warszawa 1, Poland

PORTUGAL:

Rede dos Emissores Portugueses (REP)
Rua D. Pedro V N° 7, 4°
1200 Lisboa, Portugal
Telephone: 361186

ROMANIA:

Federatia Romana de Radioamatorism (FRR)
Post Office Box 05–50
R-76100 Bucharest, Romania
Telephone: 37.05.13

SAN MARINO:

Associazione Radioamatori della Rep. di San Marino
Via delle Carrare 63
47031 San Marino
Telephone: 907851

SENEGAL:

Association des Radio-Amateurs du Senegal (ARAS)
Post Office Box 971
Dakar, Senegal

SIERRA LEONE:

Sierra Leone Amateur Radio Society (SLARS)
Post Office Box 10
Freetown, Sierra Leone
Telephone: 22246

SINGAPORE:

Singapore Amateur Radio Transmitting Society (SARTS)
G.P.O. Box 2728
Singapore 9047
Telephone: 22246

SOLOMON ISLANDS:

Solomon Islands Radio Society (SIRS)
Post Office Box 418
Honiara
Guadalcanal, Solomon Islands
Telephone: 551 Ext. 310

SOUTH AFRICA:

South African Radio League (SARL)
Post Office Box 3911
8000 Cape Town, South Africa
Telephone: 021–434443

SPAIN:

Union de Radioaficionados Espanoles (URE)
Post Office Box 220
Madrid, Spain
Telephone: (91) 274 83 97

SRI LANKA:

Radio Society of Sri Lanka (RSSL)
Post Office Box 907
Colombo, Sri Lanka
Telephone: 588718

SURINAME:

Verenigning van Radio Amateurs in Suriname (VRAS)
Post Office Box 1153
Paramaribo, Suriname
Telephone: 97788

SWAZILAND:

Radio Society of Swaziland (RSS)
Post Office Box 21
Ezulwini, Swaziland

SWEDEN:

Foreningen Sveriges Sandareamatorer (SSA)
Ostmarksgatan 43
S-123 42 Farsta, Sweden
Telephone: 08–644006

SWITZERLAND:

Union Schweizerischer Kurzwellen-Amateure (USKA)
Post Office Box 9
4511 Rumisberg, Switzerland
Telephone: 065 76 36 76

SYRIA:

Technical Institute of Radio (TIR)
Post Office Box 35
Damascus, Syria
Telephone: 717570

THAILAND:

Radio Amateur Society of Thailand (RAST)
G.P.O. Box 2008
Bangkok, Thailand
Telephone: 391–1269

TONGA:

Amateur Radio Club of Tonga (ARCOT)
c/o University S. Pacific Ext. Center
Post Office Box 278
Nuku'alofa, Tonga

TRINIDAD AND TOBAGO:

Trinidad & Tobago Amateur Radio Society (TTARS)
Post Office Box 1167
Port of Spain, Trinidad
Telephone: 637–3253

TURKEY:

Turkiye Radyo Amatorieri Cemiyeti (TRAC)
Post Office Box 109
Istanbul, Turkey

U.S.S.R.:

RadioSport Federation of U.S.S.R. (RSF)
Post Office Box 88
Moscow, U.S.S.R.
Telephone: 491–86–61

UNITED KINGDOM:

Radio Society of Great Britain (RSGB)
Alma House, Cranborne Road, Potters Bar
Herts. EN6 3JW, England
Telephone: 59015

UNITED STATES:

American Radio Relay League (ARRL)
225 Main Street
Newington, CT 06111
Telephone: (203) 666–1541

URUGUAY:

Radio Club Uruguayo (RCU)
Post Office Box 37
Montevideo, Uruguay
Telephone: 91–7742

VENEZUELA:

Radio Club Venezolano (RCV)
Post Office Box 2285
Caracas 1010A, Venezuela
Telephone: 781–77–54

WESTERN SAMOA:

Western Samoa Amateur Radio Club (WSARC)
Post Office Box 1069
Apia, Western Samoa

YUGOSLAVIA:

Savez Radio-Amatera Jugoslavije (SRJ)
Post Office Box 48
11—1 Beograd, Yugoslavia
Telephone: 11 33 22 16

ZAIRE:

Union Zairoise des Radio-Amateurs (UZRA)
Post Office Box 1459
Kinshasa 1, Zaire

ZAMBIA:

Radio Society of Zambia (RSZ)
Post Office Box 332
Kitwe, Zambia

ZIMBABWE:

Zimbabwe Amateur Radio Society (ZARS)
Post Office Box 2377
Salisbury, Zimbabwe

# · Appendix 4 ·

# Amateur Call Sign Districts

1:   Connecticut, Maine, Massachusetts, New Hampshire, Rhode Island, Vermont
2:   New Jersey, New York
3:   Delaware, District of Columbia, Maryland, Pennsylvania
4:   Alabama, Florida, Georgia, Kentucky, North Carolina, South Carolina, Tennessee, Virginia
5:   Arkansas, Louisiana, Mississippi, New Mexico, Oklahoma, Texas
6:   California
7:   Arizona, Idaho, Montana, Nevada, Oregon, Utah, Washington, Wyoming
8:   Michigan, Ohio, West Virginia
9:   Illinois, Indiana, Wisconsin
0:   Colorado, Iowa, Kansas, Minnesota, Missouri, Nebraska, North Dakota, South Dakota
KH6:   Hawaii
KL7:   Alaska
KP4:   Puerto Rico

# · Appendix 5 ·

# Sample Code Conversation (QSO)

The following is an example of code conversation that a novice might have on the air. It is given here as an indication of how we go about these code contacts. Although most code conversations (QSOs) are like this to begin, we often go on to talk about the weather, about our kids, our jobs, or anything else.

The example given is not the complete course on procedures, of course, but it is all you need to get by if you don't have any other guidelines to use.

To begin, here are a few abbreviations we use:

CQ: This is an internationally known abbreviation that means "calling any station." We use it when we first go on the air to look for someone to contact. It has about the same meaning as when you yell in the door, "Is anyone home?"

DE: This abbreviation means "from." We use it in code, for example, to say DE K2VJ which means "from K2VJ."

AR: This abbreviation means "end of transmission." It is used at the end of our comments and when we are ready to turn the conversation back to the other station for their turn to talk.

K: Sending this abbreviation at the very end of a transmission invites the other station to start transmitting. It is used together with AR; AR and K mean "I am finished. Your turn. Go ahead."

BT: This abbreviation is used like a period at the end of a sentence. There is another signal for the period, but we mostly use BT in informal contacts.

Now, putting all this together, your first novice CQ call might go like this (and I will use my call sign for convenience):

CQ CQ CQ CQ CQ DE K2VJ K2VJ K2VJ AR K

We send CQ about five times in order to catch the attention of some station that might be tuning by. If we were to send it only once, it is not likely that anyone would hear it except purely by accident.

When we send DE, we alert all stations who are listening that we are about to send our call sign. We send our call sign three times in order to give the other station plenty of chance to get it right. After all, there may be interference or, as sometimes happens, you might be a little bit nervous on that first call. You might have a problem in keeping a sweaty hand on your code key, so you send your call often enough so that anyone listening will get it right. Otherwise, no one will answer your CQ call.

There is no guarantee that someone will always come back to your call.

What do you do if no one answers? You go through the same sequence of CQs, the DE, your call sign and the AR and K. Try it a third time if no one answers. After that, either move off frequency to try a new spot on the dial or just wait a bit before making another CQ call. Or, tune around the band for someone else who is calling CQ and answer their call. Although you get better results answering other CQs than you do in calling your own, I personally feel it is best that you call CQ for your very first contact on the air. It gives you a nice memory for the years to come.

Now, let's suppose that someone answers your CQ. (We'll use my son Jim's call sign, WA2JNN). He will send, in response:

K2VJ K2VJ DE WA2JNN WA2JNN WA2JNN AR K

In other words, he sends my call sign first just like you call someone by name on the telephone when they pick it up after your ring; he sends DE to let me know he is about to send his call sign;

and he closes out with AR K to say he is finished calling me and invites me to go ahead.

Typically he will send your call sign (K2VJ) at least twice just to be sure you know you are being called and that it isn't someone else he is calling. He sends his call sign at least three times to give you enough time to get his call sign correct.

Suppose you recognize your call sign being sent in answer to your CQ, but for some reason you didn't get, or weren't sure of, the call sign of the station calling you. You ask for a repeat, and you do that with one of our international Q signals, of which we have several to fit almost every situation. In this case, the signal is QRZ? which means, "Who is calling me?" (There is a list of the more popular Q signals at the end of this appendix.)

You send:

QRZ? QRZ? DE K2VJ AR K

You should only have to send your call once because, after all, you know the other station already knows your call sign. The problem is that you don't know his or her call sign. You close with the usual AR K.

The answer to your QRZ? will probably be:

K2VJ DE WA2JNN WA2JNN WA2JNN AR K

The other station will send her or his call sign several times to give you plenty of chance to get it right. But if you don't, just go back with another QRZ? call as previously sent. Since you are among friends, it would be silly for you to quit in discouragement just because you didn't get the other station's call sign right the first or second time. Any good operator would gladly stay with you to help and encourage you.

Let me go back over what we have done so far on this very special, first on-the-air contact you are about to have.

1. You started off your call with several CQs, which is the code abbreviation for a call to any station, which means that you are looking for a contact.

2. You put DE after the string of CQs to let anyone listening know that you are about to send your call sign.

3. You sign your call sign three times.

4. You finish your CQ call with an AR and K which mean "end of my call" (or end of any transmission) and that anyone who is listening should go ahead and transmit.

5. If no one answers, you repeat the CQ call exactly like the first time.

6. If someone answers but you don't get the call sign, ask him or her to repeat the call sign by sending a QRZ? which means, "Who is calling me?"

Now, suppose you have the other station's call sign and you are ready for your first transmission. What to say? Well, that first transmission is pretty much standard. We send a signal report to let them know how well they are being received by us; we give our first names; and we also tell them where we are located. Such a transmission would go like this:

> WA2JNN DE K2VJ BT R TKS FOR THE CALL BT YOUR SIGNALS ARE RST 589 RST 589 IN EGG HARBOR NJ EGG HARBOR NJ BT NAME IS VINCE VINCE BT WA2JNN DE K2VJ AR K

If this looks like a foreign language to you on first reading, let me show you how ordinary and common it all is by giving a quick explanation.

In the first place, since we now have our call signs in order, we need send theirs and ours only once on each transmission. That's why you see WA2JNN DE K2VJ only once at this point.

I like to follow my call sign with a BT because use of BT is sort of like clearing your throat; that is, you use it like the period in a paragraph to show that something has ended and something else is about to begin. I make a lot of use of BT.

That RST 589 business has you wondering, does it? It's quite simple. It is a shortcut way we have for telling the other station how well we are hearing its transmission (or, as we more properly say, "reading" or "copying" them—they all mean the same thing, hearing, reading, receiving, and copying).

The R in RST is for "readability." How well we do, overall, read (copy or hear) the other station? It is on a scale of 1 to 5, 5 being best. When we give an R5 report, we are telling the other station that it is coming in perfectly clearly.

The S in RST is for "signal strength." How strongly are you hearing the other station? We used to judge this by ear by making an estimate of it based on a little bit of experience, but all modern radio sets now have what we call *S meters* on them which measure the signal strength of other stations right on the meter. Those S meters are scaled up to 9, which makes it easy to give a signal-strength report.

The T in RST stands for "tone" and was useful in the earlier days of amateur radio when most of us built our own transmitters. Back then, T for "tone" was very useful in telling us how neat and clean our code signal sounded over the air—or didn't. The reason it isn't of too much use anymore is due to the fact that modern commercial sets all have nice, clean signals. But we still send RST, instead of just RS, from habit, and it is just as well because if you ever get anything except a T9 report, you know something is wrong with your set that needs to be taken care of right away. The T report is in numbers from 1 to 9, 9 being the best tone.

In our sample QSO (contact), we send the RST report twice to make sure that the other station gets it right. After all, we all like to know how well our sets are being heard. We enjoy getting good signal reports. In this case, K2VJ told WA2JNN that his signal was being received 589, which means it is perfectly readable, quite strong, and excellent tone.

Now, with the RST report process understood, let me back up to the R and TKS FOR THE CALL which came before the RST. The R means, of course, "roger" or "OK." I like to make generous use of R because it is a way of saying yes, OK, I understand, I heard you—in other words, it right away assures the other station that you are doing just fine in copying them. However, don't use R if you are not copying them well.

The TKS FOR CALL means "thanks for calling me," and it is a courtesy statement many hams make. (TKS is the accepted abbreviation for "thanks.")

Right after the RST report, we send our city and state, doing it twice because many cities have some difficult spelling to them. Same with our first names.

After sending the signal report, location, and name, we follow with another BT and then we promptly turn the conversation back to the other station by sending her or his call sign once, a DE, and our call sign once, followed by the usual AR and K.

To summarize your first transmission:

1. You send each call sign one time.

2. You give a courtesy thank-you to the other station via TKS FOR CALL.

3. You use BT like periods in a sentence to separate items in your transmission. (By the way, there is a standard signal for the period which is used in formal work, but the BT signal is an informal period which many of us use instead.)

4. You send a signal report to the other station in terms of R (readability 0–5); S (signal strength 0–9); and T (tone 0–9): The RST

589 report in this example means the other station was perfectly readable, was moderately strong, and had excellent code tone. (Obviously, a reading of RST 599 is the best report we can give; it means 100 percent solid copy, very loud signal strength, and excellent quality tone.)

5. You end your first transmission with a BT, sending the other station's call sign, DE, your call sign, and the usual AR and K to invite them to transmit.

By the way, we always send the other station's call sign first, followed by ours. It's just like calling on the telephone; when someone answers, you say, "Tom, this is John." Yes, you could say, "This is John, Tom," but you can see which form is easiest to understand and least likely to be confused. We always give the other station's call sign first, followed by ours. Since DE really means "from," it makes it all the more appropriate to put our call sign after the DE (from).

Now the other station responds to your first transmission; it is likely to go just like yours:

K2VJ DE WA2JNN BT R TKS FOR CALL BT YOUR RST 599 RST 599 IN STATE COLLEGE PA STATE COLLEGE PA BT NAME IS JIM JIM BT K2VJ DE WA2JNN AR K

As you can see, it is along the same lines as your transmission was—courtesy thank you, signal report, location, name, and an invitation for the other station to transmit.

In case you miss any of the information that was sent to you, don't be bashful about asking for a repeat. For example:

WA2JNN DE K2VJ BT PLEASE REPEAT NAME BT WA2JNN DE K2VJ AR K

On the air, we frequently use ordinary abbreviations, two of which fit in here: PSE for "please," and RPT for "repeat." For example, the same transmission might have been sent:

WA2JNN DE K2VJ BT PSE RPT NAME BT WA2JNN DE K2VJ AR K

We make good use of several such abbreviations by which we shorten our words and say more with less effort in less time, particularly when operating by code. These abbreviations get to be remembered after a few uses. There is a list of the more common ones in this appendix.

In response to your request, WA2JNN will repeat his name and

then turn it right back to you. You then go into your second transmission which is also rather standard for us. You can mention your
radio set by name and model, how much power you are transmitting
with, and perhaps something about your antenna; and then you can
turn it back to the other station. It goes like this:

> WA2JNN DE K2VJ BT R JIM [We find it more personal to use
> names during a transmission.] RIG [transmitter] IS A DRAKE
> TR4 RUNNING 150 WATTS TO A HALF WAVE DIPOLE
> ANTENNA 50 FEET UP BT WA2JNN DE K2VJ AR K

(A *half-wave dipole* is the most common antenna we use—see glossary.)

The other station responds with much the same information,
then again turns it back to you.

By the way, if this is your very first QSO on the ham bands, I
think you should certainly let the other person know as much. It is
a courtesy to them, it being a rare pleasure to be someone's first
contact. First contacts set up a bond between hams, so do remember
to tell that person that this is your first contact.

Here is something else to think about. We are are not required
by law to keep a log (record) of our routine contacts. However, I
strongly suggest you do log all your contacts, particularly your first
one. I don't use the various log books that are printed for logging
purposes—I simply keep a spiral notebook on the desk, into which
I log not only the details of the contact such as date, time, frequency,
and station but also the details of our conversation. It is very impressive to contact someone on the air and say, "Oh, yes, we talked
on 40 m last year on November 3. Did you ever get a new car or are
you still driving the one that had the engine problem?" or something
like that. It makes them pay attention after that, to know someone
remembered them so well. That log book is also a treasure in future
years—take my word for it.

You might be surprised to learn that with the few transmissions
you have made in this example QSO, you have, nonetheless, been
in contact for about 20 minutes already. If you now want to end the
QSO, you send:

> WA2JNN DE K2VJ BT R TKS FOR QSO JIM BT 73 BT
> WA2JNN DE K2VJ

QSO is, of course, the contact; R is "roger," or "OK"; TKS is
"thanks"; 73, in ham radio, means "best regards" and is our commonly used way of signing off any contact. (We are forever expressing

our 73s to each other, not only over the air but also in our letters and so on.)

Actually, as you become more familiar with operating tactics, your closings will be a bit more informal and longer. It doesn't take many contacts to get the hang of it either. In fact, if you do a lot of listening over the air, you get to see how several other stations handle themselves, which is a good way to become expert at it—by listening to others.

I had mentioned earlier that you might prefer, on a later contact, to answer someone else's CQ call. Let's say that WA2JNN sends a CQ and that you, K2VJ, answer. It goes like this:

WA2JNN:  CQ CQ CQ CQ CQ DE WA2JNN WA2JNN WA2JNN AR K

K2VJ:  WA2JNN WA2JNN DE K2VJ K2VJ K2VJ AR K

WA2JNN:  K2VJ DE WA2JNN BT R TKS FOR THE CALL BT YOUR SIGNALS RST 599 599 in STATE COLLEGE PA STATE COLLEGE PA BT NAME IS JIM JIM BT K2VJ DE WA2JNN AR K

K2VJ:  WA2JNN DE K2VJ BT R JIM TKS FOR CALL BT YOUR SIGS [abbreviation for "signals"] RST 599 599 IN EGG HARBOR NJ EGG HARBOR NJ BT NAME IS VINCE VINCE BT WA2JNN DE K2VJ AR K

WA2JNN:  K2VJ DE WA2JNN BT R VINCE TKS FOR RE-PORT BT RIG IS A TENTEC CORSAIR RUN-NING 200 WATTS BT ANTENNA HR IS 3 ELEMENT KLM BEAM BT HW? K2VJ DE WA2JNN AR K [The abbreviation HW? is a con-densed form of the question, "How did you copy my transmission?" and comes from the word *how*.]

K2VJ:  WA2JNN DE K2VJ BT R JIM SOLID COPY [the expression "solid copy" is common on the bands and is used to reassure the other station that you didn't miss anything] BT RIG HR IS DRAKE TWINS AT 150 WATTS BT ANTENNA IS DIPOLE 50 FEET UP BT WA2JNN DE K2VJ AR K

At this point the conversation could run into anything either of you might care to talk about; in fact, once you are experienced at this, it turns out that you always find something to talk about on the air that is of interest to both of you. Some of the longest contacts I have ever had were those in which we continued to talk several hours after first saying we were going to sign off. Anyway, suppose this QSO is to end at this time. Here's one way to go about it:

> WA2JNN: K2VJ DE WA2JNN BT OK VINCE TKS FOR FINE QSO ES 73 BT [ES is an abbreviation for the word *and*] K2VJ DE WA2JNN AR GN

[In this last transmission, we often substitute a GN (good night), or whatever is appropriate for the time of day, in place of the K we normally send at the end of a transmission.]

> K2VJ: WA2JNN DE K2VJ R JIM PSED ["pleased"] TO WORK [contact] YOU. GL ["good luck"] ES 73 BT WA2JNN DE K2VJ AR GN

Would you like to guess how long all of this took at typical novice code speed? Would you believe a half hour? Don't be surprised— we're talking about a first or, at least, early QSO. You will get better with practice; better in both code and use of abbreviations. It takes about a month on the air to get into the hang of things, maybe less.

It really isn't at all important what you talk about in those first few contacts. What is important is that you get the hang of things, that you build up the confidence and the talent to move up the code ladder to where you can run conversations like this in about 10 minutes, leaving plenty of time for more personal talk.

## ABBREVIATIONS USED IN CODE WORK

The following is a list of the more common abbreviations we use over the air. These are used in ordinary conversation in place of spelling out everyday words. You will see how the abbreviations are taken from the word in some cases, but in other cases the abbreviations may not be apparent. The origin of these abbreviations is all in the history of radio.

ABT:   about
AGN:   again
BCNU:   be seeing you
B4:   before
C:   yes (from "si")
CLG:   calling

CUL:  see you later
DX:  distance
ES:  and
FB:  fine business; good
FREQ:  frequency
GA:  go ahead; good afternoon
GE:  good evening
GL:  good luck
GM:  good morning
GN:  good night
GND:  ground
GUD:  good
HI:  code laughter
HR:  hear; here
HV:  have
HW:  how
NR:  number
NW:  now
OM:  old man
OPR:  operator
OT:  old-timer
PSE:  please
R:  roger; OK
RCVR:  receiver
RPT:  repeat
RX:  receiver
SIG:  signal
SKED:  schedule
SRI:  sorry
TFC:  traffic
TKS:  thanks
TMW:  tomorrow
TT:  that
TU:  thank you
TX:  transmitter
UR:  your
VY:  very
WX:  weather
XMTR:  transmitter
XYL:  wife
YF:  wife
YL:  young lady

73: best regards
88: love and kisses

# INTERNATIONAL Q SIGNALS

The International Q Signals are a set of abbreviations for International Morse Code use. These abbreviations have been adopted by all countries as a way by which telegraphers can communicate with each other despite not knowing the other's language.

Prior to the sinking in 1912 of the British ship *Titanic* (when it struck an iceberg on its first trip from New York to England and sank with 1500 of its 2200 passengers), the world of international wireless telegraphy was very confusing because there were no commonly accepted abbreviations.

The *Titanic* disaster got everyone's thinking turned in the right direction. The London International Telecommunications Conference in 1912 agreed not only to adopt the International Q Signals but also to use SOS as the international distress signal. It had to be. By 1912 there were already 500 coast wireless stations and nearly 3000 shipboard wireless stations. In order for these ships to communicate with the many shore stations around the world, it was necessary to set up a series of abbreviations that would have the same meaning in every language—the abbreviations which we refer to as the International Q Signals.

At first there were 39 such Q signals (abbreviations); many more have since been added, some made especially by and for the radio amateurs.

It wasn't long after the International Q Signals were adopted for commercial telegraphy that the radio amateurs began making use of them simply because they turned out to be such an effective way to talk to radio amateurs of other countries without knowing their languages.

There are many more signals than the ones listed here; only the more common ones are given. These Q signals are in two forms, statements and questions, as demonstrated by the first signal, as follows:

QRL: I am busy. (or   QRL?:   Are you busy?)
QRM:  I am being interfered with (by other stations).
QRN:  I am troubled by static (natural atmospheric noise).
QRP:  Decrease power (also used to mean you are using low power).

QRS:  Send more slowly.
QRT:  Stop sending.
QRX:  I will call you again at (   ) hours. (Also used to mean, "Wait.")
QRZ:  You are being called by.
QSL:  I acknowledge receipt.
QSO:  I am communicating with (   ).
QSY:  Change frequency to (   ).
QTH:  My location is (   ).

## PHONETICS

We use *phonetics* on the air, particularly when transmitting by voice, to assure that letters we mention are properly understood. Phonetics are common word substitutions for letters. For example, my call sign (K2VJ) expressed by voice on the air would be given as "Kilo Two Victor Juliette."

A:  Alpha
B:  Bravo
C:  Charlie
D:  Delta
E:  Echo
F:  Foxtrot
G:  Golf
H:  Hotel
I:  India
J:  Juliette
K:  Kilo
L:  Lima
M:  Mike
N:  November
O:  Oscar
P:  Papa
Q:  Quebec
R:  Romeo
S:  Sierra
T:  Tango
U:  Uniform
V:  Victor
W:  Whiskey
X:  X-ray

Y:   Yankee
Z:   Zulu

In Chap. 18 on the novice license exam, several sample questions were given. These questions were the actual questions from the bank of 200 questions from which 20 are taken to make up a novice exam. Here are a few more questions from the section on operating procedures.

1. What does the S in the RST signal report mean?
2. What does the R in the RST signal report mean?
3. What does the T in the RST signal report mean?
4. What is the difference between the telegraphy abbreviations CQ and QRZ?
5. What is the meaning of the telegraphy abbreviations DE, R, AR, 73, QRS, QTH, QSL, QRM, and QRN?

Your exam will very likely have one of these questions in it. How well did you do with them without thumbing back through this appendix?

I think you get the idea that the novice exam isn't going to be too difficult for you, now that you have read this book. Good luck!

# · Appendix 6 ·

# The Amateur Radio Service

The formal declaration of basis and purpose to the amateur radio service, as carried in the Federal Communications Commission's documents is:

1. To recognize and enhance the value of the amateur radio service to the public as a voluntary, non-commercial communications service, particularly with respect to providing emergency communications.

2. To continue and extend the amateur radio operators' proven ability to contribute to the advancement of the art.

3. To encourage and improve the amateur radio service by providing for advancing skills in both the communication and technical phases.

4. To expand the existing reservoir within the amateur radio service of trained operators, technicians, and electronics experts.

5. To continue and extend the radio amateurs' unique ability to enhance international goodwill.